# Central Neural States
# Relating Sex and Pain

**Advances in Systems Neuroscience and Behavioral Physiology**

Jay Schulkin, Series Editor

# Central Neural States Relating Sex and Pain

RICHARD J. BODNAR

KATHRYN COMMONS

AND

DONALD W. PFAFF

The Johns Hopkins University Press
Baltimore and London

The Johns Hopkins University Press
2715 North Charles Street
Baltimore, Maryland 21218-4363
www.press.jhu.edu

Library of Congress Cataloging-in-Publication Data

Bodnar, Richard J., 1946–
   Central neural states relating sex and pain / Richard J.
Bodnar, Kathryn Commons, and Donald W. Pfaff.
     p.   cm. — (Advances in systems neuroscience and
behavioral physiology)
Includes bibliographical references and index.
   ISBN 0-8018-6827-0
   1. Neuroendocrinology.  2. Sex (Biology)  3. Pain.  4. Opioid
peptides.  5. Estrogen.  I. Commons, Kathryn, 1967–  II. Pfaff,
Donald W., 1939–  III. Title.  IV. Series.
   QP356.4.B64 2002
   612.8—dc21

2001001743

A catalog record for this book is available from the British Library.

# Contents

# Contents

# Series Foreword

This is an exciting time for molecular biology. The complete sequence of the human genome has now been mapped, and each week new genes are identified. Accomplishing the huge task of understanding the human genome, however, will require physiological and behavioral analyses to show us what the genetic information ultimately means for the functioning and malfunctioning of the body and brain. In this great age of molecular biology, systems neuroscience, and whole-body physiology —the subjects of this series—are fundamental. Now more than ever, there is a need for integrative studies.

It is also fitting that a series on integrative biology be published by Johns Hopkins University Press. Curt Richter (1919–1988), a psychobiologist in the Department of Psychiatry at Johns Hopkins for many years, was the progenitor of this field of inquiry. He studied such diverse subjects as nutritional choice in behavioral regulation of the internal milieu; psychophysics in animals and humans; bait shyness, which was a precursor to a specialized form of learning (learned taste aversion); and learned helplessness, which he called hopelessness. He even demonstrated that chronic fear or worry had real consequences for adrenal function. Well before others, Richter understood that the brain is the common effector system in the organization of behavior and that an interplay of behavior and physiology is integral to maintaining the internal milieu.

Richter's work followed and expanded on the tradition of Claude Bernard and Walter Cannon by introducing behavior that serves in the defense of the internal milieu. Now we expand on the notion of homeostasis regulation to account for how the central nervous system plays such a vital role in regulatory events, including gene expression.

*Central Neural States Relating Sex and Pain*, by Richard J. Bodnar, Kathryn Commons, and Donald W. Pfaff, launches the series. In this book, we appreciate examples of success in reasoning back and forth between neuronal systems, gene expression, and reproductive behaviors.

This book series is devoted to broad-based biological inquiry and draws on two rapidly expanding arenas of knowledge: the molecular and the cognitive. The series will encompass topics relating to behavioral and biological regulation of the internal milieu.

Jay Schulkin
Department of Physiology and Biophysics,
Georgetown University School of Medicine
American College of Obstetricians and Gynecologists
Clinical Neuroendocrinology Branch,
National Institute of Mental Health

# Preface

This book was born of an opportunity and desire to unite two major sets of neurobiological data, both of which individually are of basic experimental and practical clinical importance. Steroid hormone effects on behavior have yielded clear neural and genetic explanations. Here, we propose them to be proof of underlying, altered motivational states. Neural controls over responses to pain have provided many brilliant neurochemical, neuropharmacological, and genetic insights into behaviors of great clinical significance. Biological thinking and new experimental results have allowed us to consider them together.

At a high level of discourse, Dr. David Rubinow, of the National Institute of Mental Health, and Dr. David Edwards, professor at Emory University, gave us tremendous encouragement, useful suggestions, and important criticisms. The book could not have reached its final level without their generous and collegial support. For feedback on chapters I and II, inputs from Professors Case Vanderwolf (University of Western Ontario), Jon Horvitz (Department of Psychology, Columbia University), Jay Schulkin (Georgetown University School of Medicine and American College of Obstetrics and Gynecology), Yasuo Sakuma (Professor of Physiology, Nippon University, Tokyo), and Carol Dudley (Department of Physiology, Southwestern Medical School) are invaluable. The neurochemistry and neurophysiology of pain and its abatement comprise a major theme in chapter III. Here, Drs. Howard Fields (Professor of Neurology, UCSF), Alan Gintzler (Professor of Biochemistry, SUNY Downstate College of Medicine), Eric Simon (Professor of Psychiatry and Pharmacology, New York University School of Medicine), Mary Heinricher (Division of Neurosurgery, Oregon Health Sciences University), Herbert Proudfit (Professor of Pharmacology, University of Iowa College of Medicine), Richard Bandler (Professor of Anatomy, University of South Wales), Martin Kavaliers (University of Western Ontario), and Jeffrey Mogil (University of Illinois) provided excellent advice and suggestions. Dr. Alan Herbison (Babraham, Cambridge, U.K.) gave us valuable neuroendocrine insights. The small clinical sections of the book have benefited from information supplied by Drs. Robert Michels (University Professor, Cornell Medical School) and Margaret Moline (Department of Psychiatry, Cornell Medical School).

We are all extremely grateful to our laboratory coworkers, friends, and close relatives for suffering through the preparation of this book with us. Insofar as the studies relating to analgesic processes are concerned, I (RJB) am totally indebted to the perseverance and hard work of present and former graduate students in my laboratory who performed these studies, including Diana Badillo-Martinez (Danbury Hospital); Pam Butler (Nathan Kline Research Institute); Anita Islam (Target Health Inc.); Karen Kepler; Ben Kest (College of Staten Island); Jackie Kiefel (Emory University); Annette Kirchgessner (SUNY Health Science Center at Brooklyn); Jeff Kordower (Rush-Presbyterian Medical Center); Elisse Kramer (Long Island Jewish Medical Center); Ed Lubin (Tufts University); Zoran Pavlovic; Judith Robertson; Maria-Teresa Romero (State University of New York at Binghamton); Grace Rossi (C.W. Post College and Memorial Sloan-Kettering Cancer Center); Randi Shane; Don Simone (University of Minnesota School of Medicine); Ellen Sperber (Mercy College); and Marcello Spinella (Richard Stockton State College). I must also acknowledge my mother, Irene; my sister, Ann; my graduate mentors, Steve Ellman and Sol Steiner; my long-standing scientific "brother," Bob Ackermann; my postdoctoral collaborators, Dennis Kelly and Murray Glusman; and especially my long-standing collaborator of almost 20 years, Gavril Pasternak. Most of all, I continue to draw inspiration, unstinting support, and unparalleled patience from my sons, Ben and Nick, and especially my wife, Carol Greenman.

Lucy Frank, of the Rockefeller University, helped put this book together. This complex endeavor involved dealing with many references and figures from three different authors, and the book owes its high state of organization to her.

# List of Abbreviations

| | |
|---|---|
| AS ODN | antisense oligodeoxynucleotide |
| β-FNA | β-funaltrexamine |
| CCWS | continuous cold-water swim |
| CTOP | $Cys^2$-$Tyr^3$-$Orn^5$-$Pen^7$ |
| DADL | $D$-$Ala^2$, $D$-$Leu^5$-enkephalin |
| DAMGO | $D$-$Ala^2$, $Met$-$Phe^4$, $Gly$-$ol^5$-enkephalin |
| DELT | $D$-$Ala^2$, $Glu^4$-deltorphin |
| DPDPE | $D$-$Pen^2$, $D$-$Pen^5$-enkephalin |
| DRN | dorsal raphe nucleus |
| DSLET | $D$-$Ser^2$, $Leu^5$-enkephalin |
| EAA | excitatory amino acid |
| ICWS | intermittent cold-water swim |
| LC | locus coeruleus |
| M6G | morphine-6β-glucuronide |
| MPOA | medial preoptic area |
| Nacc | nucleus accumbens |
| NalBzOH | naloxone benzoylhydrazone |
| NBNI | nor-binaltorphamine |
| NRGC | nucleus reticularis gigantocellularis |
| NRM | nucleus raphe magnus |
| NTI | naltrindole |
| NTS | nucleus tractus solitarius |
| ORL-1 | orphanin-opioid receptor |
| OVX | ovariectomized |
| PAG | periaqueductal gray |
| POA | preoptic area |
| POMC | proopiomelanocortin |
| PPE | preproenkephalin |
| PVN | paraventricular nucleus |
| RVM | rostral ventromedial medulla |
| SN | substantia nigra |

| SPA | stimulation-produced analgesia |
| vlPAG | ventrolateral periaqueductal gray |
| VMH | ventromedial nucleus of the hypothalamus |
| VTA | ventral tegmental area |

# Central Neural States
# Relating Sex and Pain

# I

# *Requirement for Motivational State Concepts*

This book argues that central nervous system (CNS) states of arousal and motivation can be heightened by sexually related stimuli as well as by painful stimuli. In turn, these CNS states influence the production of both responses to sexual behavior (to facilitate them) and responses to pain (to reduce them). Close relations between pain and sex are fostered by an overlap between the two sets of neuroanatomical pathways and mechanisms. As a result, damping of pain responses permits sexual responses to proceed normally.

Will neurobiologists be limited to explaining behavioral responses one by one, or will neuroscience advance quickly enough that we can gain insight into entire classes of logically related responses? Some biologists fear any formulations that group behavioral responses logically, and they want to limit themselves to individual, concrete, operationally defined experimental situations. We can do better. Suppose physicists had been limited in such a concrete manner. They would have had to construct separate descriptions for the velocity of a Coca-Cola bottle dropped from the Leaning Tower of Pisa, a root beer bottle, a Pepsi bottle, and so forth. No lawful description of acceleration of objects due to gravitational force would be allowed to emerge. It is an intellectual necessity to imagine and construct groups of behavioral phenomena that are subject to lawful neural explanation. Tying together behaviors by concepts such as motivation and arousal will be required for fast scientific progress. Doing this will also increase the possibility of linking discoveries in the mouse to problems in the clinic.

Likewise, it is a big advantage to go beyond the single expressed re-

1

sponse, that is, by considering states of the CNS that predispose the animal or human toward certain groups of behaviors and away from others. Indeed, it is a physical necessity that a particular state of CNS activity precede a specific behavior because, after all, sets of appropriate neurons produce that behavior. Most behavioral responses above the level of the reflex do not happen instantaneously. They result from coordinated relationships among neuronal subsystems that produce states favoring their occurrence on presentation of the adequate stimulus. Some CNS states deal with brain arousal, the motivation of behavior, and mood. Those are the types of CNS states we deal with here.

Why CNS states triggered by sex and pain? Because they exhibit intriguing parallels and contrasts. Both depend largely on somatosensory stimuli. Both impact broad ranges of behavioral responses. Neither is homeostatic in the sense of food intake, water intake, or temperature regulation. Both are of great clinical importance. Yet, they carry opposite valences: leading to approach responses for sexual stimuli and avoidance responses for pain.

We build on well-described neural circuitry and genetic mechanisms for sexual behaviors (Pfaff, 1999a) and two scientific generations of work on brain mechanisms controlling responses to pain. How can we conceive of the former as due to a specific sexual motivational state? Can we relate it to the latter, the general arousal by and specific response to pain?

There has been considerable success in explaining a natural mammalian social behavior, the female-typical mating behavior lordosis, in neural and genetic terms (Pfaff, 1999a). Lordosis behavior comprises an extreme vertebral dorsiflexion biologically crucial for reproduction because it allows fertilization. For genomic analysis it is advantageous because it depends on priming with estrogenic hormones, whose facilitatory influences can then be amplified by progestins. These classes of hormones work, in part, through their cognate nuclear receptors, which are, in fact, genetic transcription factors. The neural circuit for the behavior starts with ascending systems carrying somatosensory information from lumbar spinal cord levels up to the brain stem reticular formation and midbrain periaqueductal gray (PAG). Controlling the circuit are the neurons of the ventromedial nucleus of the hypothalamus (VMH), whose outputs activate systems in the midbrain PAG that then, through a hierarchy of descending systems, allow lordosis behavior to be executed. The discovery that the neuropeptide gonadotropin-releasing hormone could facilitate lordosis behavior (Moss and McCann, 1973; Pfaff,

1973) helped to explain the beautiful coordination between reproductive behavior and ovulation. Strategic advantages in the analysis of this behavior include its relatively simple stimuli and response properties, its dependence on the simple chemicals that are steroid hormones, and the fact that their receptors' ability to act as transcription factors brings the problem into the realm of molecular genetics.

Because of the success in showing concrete mechanisms for this behavior, we are emboldened to consider its underlying sexual motivation (chapter I). While some mechanisms of motivation surely are specific to the drive toward reproduction, others are broadly distributed neuroanatomically and generalized across several motivational states. Virtually all theories and data classify the more generalized motivational influences with mechanisms of "activation" and "arousal" (chapter II). In turn, ascending arousal systems themselves contain broadly generalized features (connected with sleeping and waking) as well as more particular features related to sensory alertness, motor activity, and emotional reactivity (chapter II). Thus, in this book we have the opportunity to address and help clarify some of the most fundamental issues of CNS physiology and chemistry. We deal with the neural systems that keep us as sentient and reactive as human beings. Here, the domains of sex and pain are approached directly.

But sexual arousal and pain signaling hardly comprise the entirety of the mechanisms serving reproductive behaviors and defensive behaviors, respectively. For example, when the VMH at the top of the circuit for controlling lordosis behavior (Pfaff, 1980) has been activated, what then? The sex hormone dependence of gene activation for the opioid peptide enkephalin (Romano et al., 1988; Romano, Krust, and Pfaff 1989; Romano et al. 1989; Lauber et al., 1990; Priest, Borsook, and Pfaff, 1996) tipped us off to the fact that moderating the female's reaction to otherwise noxious and strong somatosensory stimuli may be an essential part of reproductive behavior in some animals. Control of pain and reactions thereto are thus drawn into elementary neural and genetic mechanisms related to reproduction. Indeed, relations between pain and sex have a long clinical history (chapter III).

The final set of arguments in this book (chapter IV) summarizes certain modern approaches to the influences of genes on mammalian behaviors. It also attempts to weave together the longest and strongest threads of mechanistic arguments presented, in order to envision how relations between sex and pain are manifest in states of elevated general arousal and specific sexually motivated behaviors.

## A. Physics, Brain, and Behavior

Many behavioral neuroscientists have followed an intellectual path that has more in common with physics than with biology. Even in the early nineteenth century, the physicist Alexander von Humboldt celebrated "the deep feeling for unity in nature." His sympathy finds strong resonance with the attitude of Goethe, who strove, in the words of Gerald Holton "toward a unified fundamental understanding of nature," with the ambition of achieving "a single, coherent world picture (*Weltbild*)." As quoted by Holton, Einstein said that the supreme task of the physicist is "to arrive at those universal elementary laws from which the cosmos can be built up." By way of comparison, great theories intended to explain the biological basis of behavior in the last century, for example, in the thinking of Clark Hull, Neal Miller, Donald Hebb, and many others, strove to achieve the very simplest explanations of complex behavioral phenomena. Far from reveling in complexity for its own sake, in diversity among cells, individuals, or species, this tradition in the physiological approach to behavior seeks laws that have a universal character. While during the last few decades certain biologists have tended to seek "examples," behavioral neuroscientists often have struggled to enunciate and confirm laws that could have widespread validity.

In that spirit, animal behaviorists and neurophysiologists trying to explain alterations in behavior in the face of a constant environment and uniform stimuli came up with the concept of central motivational states (Bolles, 1961; Young, 1961). To explain this concept, we use an analogy between the logic of mathematical terms in Newton's Second Law appealing to gravitational force, and the logical requirement for motivational concepts in the explanation of behavioral plasticity (Fig. 1). No one has ever "seen" gravity. It is required to account for the relation between mass ($\mathbf{m}$) and acceleration ($\mathbf{a}$). Likewise, one infers the requirement for and amount of motivation. It is required to explain quantitative alterations of response in the face of a constant stimulus and context.

A different kind of physical analogy would be to picture behavior as a vector. All vectors are characterized by length and direction. If the very strength of neural and genetic influences leading to a behavioral response, the motivation, is similar to the "length of a vector," then the selectivity of a behavioral response to a particular stimulus would be parallel to the "direction of the vector."

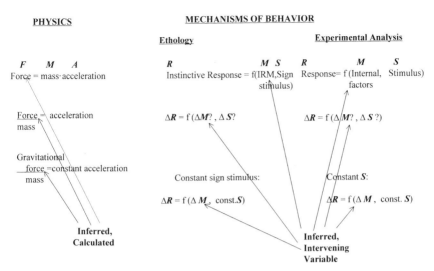

**Fig. 1.** Consider the rough analogy between the logical status of motivational concepts as mechanisms in behavior and the logical status of the gravitational concept in physics. In physics, forces are calculated according to Newton's Second Law in a manner that explains the relationship between mass and acceleration. No one has ever "seen" gravity. Where gravitational force is involved, it has been inferred. The inference of a gravitational force is required to explain the constant acceleration (32 ft/sec/sec) due to gravity. Likewise, for explaining changes in the occurrence of instinctive responses, ethologists such as Tinbergen (1951) were forced to assume that an innate releasing mechanism (IRM) stood between the releasing "sign stimulus" and the motor neurons governing the response. Changes in this IRM had to be inferred to explain changes in the releasing value of specific, constant sign stimuli. Modern studies of reproductive behavior solve Tinbergen's problem in neural and genetic terms (e.g., Pfaff, 1999a). Finally, in the scientific tradition of the experimental analysis of behavior, what would explain changes in behavioral responses to a constant stimulus? An intervening variable had to be inferred. Where clear, measurable, and manipulable biological variables such as hormones and changes in nutrition or hydration could be drawn into the explanation, the logically required, inferred intervening variable could be called "motivation."

## B. Motivation in Its Generalized and Specific Aspects

The first behavioral theorist who, forcefully called attention to the simplifying power of the notion of a central motive state was Morgan (1943), who suggested that such states, aroused by stimulus, hormonal, or experiential factors, could account for considerable numbers of behavioral responses. That is, to account for variations in behavioral response to a constant stimulus in a constant environment, variations in central motive states—most clearly reflecting biological needs—would be physiologically sensible and theoretically satisfying. As summarized by Bolles (1961), central motive states persist in time, they heighten general bodily activity, they evoke specific forms of behavior effective in satisfying said bodily need, and they prepare the organism for appropriate consummatory behaviors when environmental conditions allow. As such, they represented internal "stimuli" to activity, and their elucidation has dominated large segments of subsequent neurobiological research. Even at that time, the concept of central motive state was consistent with the evidence then available on male sexual behavior, for which Beach (1942) postulated "central excitatory mechanisms." Theorists of the relationships between motivation and emotion, for example (Young, 1961), also recognized the explanatory power of "central motive states." Knowing that such neurophysiological states might have an element of nonspecificity in their neuroanatomical representations—including the brain stem reticular formation and the hypothalamus—theorists nevertheless emphasized that central motive states could account for the arousal of behavior, the maintenance of that behavior, and the satiation or termination of such behavior (Young, 1961).

Encompassed by the notion of a central motive state was the set of neurophysiological properties exhibited by an ascending reticular activating system. Hebb (1955) incorporated the states of arousal consequent to this ascending reticular system in his theory of motivation. In Hebb's view, the explanation of motivational states was central to the general problem of understanding behavior. Increasing levels of arousal would contribute toward central neural states that not only would drive consummatory responses but would also reach an optimal level for the learning of new responses. In his behavioral theorizing, Hebb was obviously influenced by the neurobiological work of Lindsley and colleagues (Lindsley 1951, Lindsley, Bowden, and Magoun 1949, Lindsley et al. 1950) and Moruzzi and Magoun (Moruzzi and Magoun, 1949; Magoun, 1958), which opened up the subject of the neurophysiological underpinnings of arousal (see chapter II).

Reinforcing these views were the results of Valenstein, Cox, and Kakdewski (1970) dealing with the behavioral results of the electrical stimulation of the hypothalamus. They challenged the notion that stimulation of discrete hypothalamic areas would yield behaviors that were limited to the expression of individual, specific drive states. From their results, it did not appear true that there was an isolated circuit devoted to stimulus-bound eating, another devoted to stimulus-bound drinking, and so forth. Instead, it appeared that, far from the activation of only one specific behavior pattern by hypothalamic stimulation, a group of responses related to a common state might be elicited by electrical stimulation. This kind of electrophysiological work also lent credence to the notion of a central motive state.

Neither the breadth nor the persistence of the concept of central motive states should be underestimated. Proposed mechanisms were studied and reviewed in detail (Stellar, 1954; Gallistel, 1980; Stricker and Zigmond, 1984). Thus, the potential explanatory powers of central motive states have been explored for almost sixty years. Nor are they limited to muscular and behavioral manifestations. To understand the hypothalamic control of hunger motivation, autonomic states have been invoked (Powley, 1977). Nor is the concept limited to vertebrates. The feeding behavior of a mollusk, *Aplysia,* associated with an arousal state characterized by a distinct behavioral profile, depends centrally on the electrical activity of a particular neuron responsible for the alteration of body posture, biting, and cardiovascular changes. Food stimuli, which evoke food-related arousal responses in this animal, produce prolonged excitation of this neuron (e.g., Teyke, Weiss, and Kupferman, 1990; Teyke et al. 1997). Thus, in this lower animal, as in mammals, particular constellations of neuronal activity may best be summarized for the purpose of behavioral explanations and neurophysiological investigations as representing central motive states.

A motivation theorist whose formulations most comprehensively dealt with the concept of central motive state was Dalbir Bindra, a professor at McGill University in Montreal. While he concentrated on the effects of motivational changes as registered in general activity, he was also interested in the manner by which general motivational changes could facilitate specific instrumental responses (Bindra, 1968). It appeared that the logical, systematic, and quantitative ambitions of his theories derived from the conceptual advances of Clark Hull and colleagues at Yale (Hull, 1943, 1952). Bindra's definition of *central motive states* was of a functional neural change generated by the interaction between the neural representations of bodily states and appropriate incentive objects (Bindra, 1969). As such, a central motive state was pictured as fostering both exploratory and instrumental (appetitive) behavioral acts and consummatory acts. For Bindra (1969), central motive states would influence behavior by facilitating or inhibiting specific sets of sensory-motor mechanisms and would activate exploratory and appetitive responses at lower thresholds than those for consummatory responses.

European ethologists, as well, recognized the importance of internal states for the explanation of behaviors that did not follow slavishly on the presentation of easily identified external stimuli. Tinbergen (1951) discussed internal factors responsible for changes in behavioral responses that did not depend simply on the releasing value of sensory stimuli. He included hormones among those factors but also referred to other types of central neural factors. His concepts were more complex than the straightforward "hydraulic" model of motivation promulgated by the equally famous ethologist Konrad Lorenz (1950) but made essentially the same point. In a modern vein, Robert Hinde (1960, 1970) affirmed the importance of motivational concepts and directed considerable attention toward the neurophysiological experiments necessary to discern their mechanisms. However, he resisted the simplifying ideas related to a single "central excitatory state" deriving from Lorenz's approach, emphasizing instead the diversity of central neural changes that could account for a variety of behavioral responses (Hinde, 1960).

Where in the nervous system might these motivational changes be generated? Eliot Stellar (1954), a professor at the University of Pennsylvania, was the first to postulate in a clear fashion that the amplitude and duration of motivated behaviors were a direct function of neuronal excitation in particular cell groups in the hypothalamus. In retrospect, with decades of additional neurophysiological, neuroanatomical, and behavioral work behind us, it is easier to see how hypothalamic mechanisms of motivation are integrated with distributed ascending and descending systems in the brain stem and even in the basal forebrain. Already in the 1960s, Neal Miller and colleagues (Miller, 1965) were using neurochemical techniques to characterize the nature of central neural mechanisms for motivation. In all cases, Miller's (1959) approach required the convergence of different types of experimental evidence to infer the existence of a specific form of motivational state.

In this book, we put forth the thesis that certain central motivational states are important for the responses of animals and human beings to sexually relevant stimuli and to pain (Fig. 2). As neural and genetic reductionistic techniques came into vogue, some neuroscientists, not educated in the experimental analysis of behavior, thought that new methodologies would disfavor universal behavioral concepts. Nothing could be further from the truth. For states related to arousal (chapter II) and analgesia (chapter III), behavioral responses to sexually relevant stimuli and painful stimuli share the activation of certain neuronal systems and, in doing so, help explain bodies of experimental and clinical data in an-

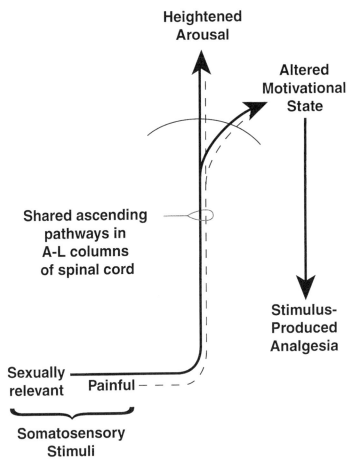

**Fig. 2.** Painful somatosensory stimuli and sexually relevant somatosensory stimuli share overlapping ascending pathways in the anterolateral columns of the spinal cord. Their actions in the brain stem both heighten arousal and induce stimulus-produced analgesia.

imals and humans, respectively. Motivational state concepts are powerful theoretically because they can address, at once, multiple domains of behavioral responses.

## C. Motivation for Females to Seek Males

For the simplest form of mammalian reproductive behavior, lordosis behavior, and for the simplest definition of motivation—an intervening

variable logically required to explain the occurrence of lordosis following estrogens and progesterone—many neural and genomic mechanisms have been explained thoroughly (Pfaff, 1999a). In this restricted
sense, therefore, the motivation problem has been "solved." Far beyond
this simple, consummatory behavior, however, females and males en-
gage in numerous communicative and exploratory behaviors, ranging
from simple affiliation through intricate courtships. Much experimental evidence shows that there exists a motivation for female animals to
seek males, and vice versa. Here we argue that the aroused female rat,
with heightened electrical activity and neurochemical activation in specific regions of the lateral preoptic area (POA), will show a tremendous
increase in locomotion. In the vicinity of males, this can result in approaches toward the males and even courtship.

    In controlled laboratory studies with rats and mice, the willingness of
females to approach males is beyond doubt. In the older literature, it was
not only demonstrated that female rats would approach males, but that
the natural limitations on such approach responses were also spelled
out. Peirce and Nuttall (1961) gave female rats free access compartments
with sexually active male rats (mating compartments) as well as to ones
without males (escape compartments) in order to discover how the females would "pace" their contacts with the males. In an orderly fashion,
all females paced their contacts with male rats according to the intensity
of their immediately preceding contacts. Following the least intrusive encounters with males—mounts without intromissions—females went to
the escape compartment only 37 percent of the time. After intromissions, they left the mating compartments 95 percent of the time, and following ejaculation by a male, 100 percent of the time. Further, the
amount of time during which females prevented access by a male was
also an orderly function of the intensity of the preceding somatosensory
contact: after no contact, only 7 s; after a simple mount, 27 s; after an in-
tromission, 60 s; and after an ejaculation, 218 s. Peirce and Nuttall
(1961) concluded that the motivation for the female rat to approach the
male (or to allow approach by him) was conditioned by those circumstances under which contact with the male would be aversive.

    Another index of the willingness of the female rat to approach the
male rat is the performance of an arbitrarily chosen operant response
by the female to allow entry by the male. Bermant (1961) showed that
female rats would perform an operant response to allow entry by the
male and, similar to the study of Peirce and Nuttall (1961), demonstrated
that the latency to the performance of that operant response varied

according to the type of previous contact with the male. Delays before performance of the operant response were shortest following simple mounts, compared with the longer delays following intromissions and the still longer delays following ejaculations. To analyze the nature of the somatosensory stimulation that accounts for the delays of response by the female, Bermant and Westbrook (1966) found that, following ejaculations, formation of the vaginal plug by the ejaculate was an important factor. When the formation of such a plug was prevented, response intervals were shorter. In addition, local treatment of the epithelium with an anesthetic reduced delays of response. Finally, when contacted by males that could not intromit, females showed short delays of response to readmit males, even throughout long tests. In summary, female rats demonstrate a motivation to approach males or allow males to approach them, but this may be limited, at least, by a history of recent, intense, somatosensory contact.

By using modern quantitative approaches to the experimental analysis of behavior and corresponding data-collection systems, it has been possible for researchers to replicate and extend these early observations (Matthews et al., 1997). The use of operants such as nose poking to break a beam of light, bar pressing, and gentle nose pressing against a sensitive button all showed that females would learn such arbitrarily chosen responses to gain access to tethered males (Fig. 3). This conclusion was enforced throughout a variety of tests with female rats and mice (Fig. 4). Because, in this type of assay, females would work to gain access to other females as well as males, and because the results appeared independent of gonadal state, we interpreted the findings as reflecting a general tendency toward affiliation (Matthews et al., 1999). For example, figures 3 and 4 are representative of years of work yielding cumulative records produced by female rats and mice performing a learned nose-press response to gain access to either another female or male. It appears, therefore, that among the different types of motivation that could bring females and males together, a nonspecific tendency toward affiliation could precede specifically sexual motivation.

Estradiol (E) has been shown to elevate specifically sexual motivation (see above). A large body of evidence first reviewed in Pfaff (1982) has since been augmented with modern techniques. Evidence gathered in many laboratories over a long period of time suggests strongly that estrogenic hormones can increase the tendency for female rats to approach males. For example, Warner (1927) used an apparatus in which the test rat had to cross a chamber, the floor of which was an electrified

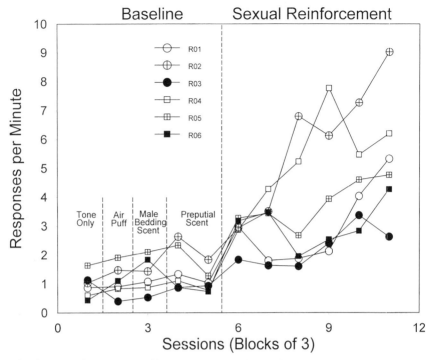

**Fig. 3.** Female rats were willing to perform a learned operant response, nose poking to break a light beam, to gain access to a male. Operant response rates for a control, a conditioned stimulus, a tone, a simple air puff (as control for pheromonal stimuli), the scent of male rat bedding, or odors derived from male preputial glands all were quite low. On the other hand, when performance of the operant response resulted in actual access to a stud male rat, rates were significantly higher. (From Matthews et al., 1999.)

grid, in order to enter a compartment housing a male. In female rats, the number of grid crossings was by far greatest during the period of the estrous cycle occasioned by a cornified vaginal smear, which reflects high estrogen levels. For a fixed test duration, during the metaestrous (low estrogen) types of smears, none of the 32 female rats crossed more than five times. In dramatic contrast, during a period of cornified smears (high estrogens), 18 of 21 females crossed more than five times. As an experimental control, when there was no male in the test compartment beyond the electrified grid, the frequency of crossings was only about one-third of that seen when the male was present.

Nissen (1929) used the Columbia Obstruction Apparatus, in which animals had to cross an electrified grid to get to an "incentive compartment." A test rat was allowed to make four crossings from the entrance to the incentive box with no shock. This established the incentive. On the last preliminary crossing, the circuit was closed while the animal was on the grid but after the door leading back to the entrance compartment had been closed, so that the test rat was forced to pass from a place where it was being electrically shocked to the incentive compartment. Then, the experimental trials began. Nissen's results showed clearly that OVX,

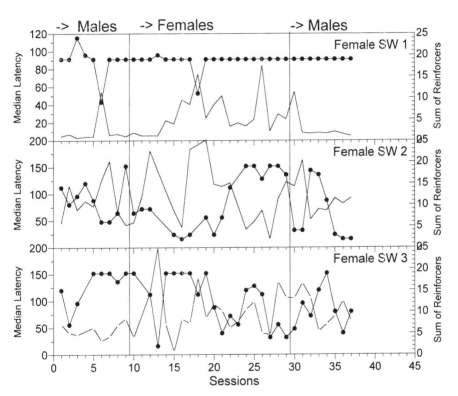

**Fig. 4.** Female OVX (ovarectomized) mice were willing to perform a learned operant response, nose pressing against a sensitive button, to gain access to either stimulus males or stimulus females. Examination of individual records of either response latency (o—o) or the total number of reinforcers per session (—●—) indicates approximately equivalent strengths of response to obtain access to either males or females (Matthews et al., unpublished data). We hypothesize that some response situations lead to behavior based on a general affiliative motivation, which, in turn, produces situations that lead to specifically sexual motivation.

uninjected control female rats and cycling diestrous females hardly ever crossed the grid to get to a male. However, cycling female rats with cornified smears (indicating high estrogen levels) crossed the grid significantly more often, and, likewise, OVX animals injected with placental extracts crossed very frequently. Nissen inferred that ovarian hormones increased the female rat's motivation to approach the male.

Jenkins (1928) analyzed the nature of the requirement for the incentive animal in this type of test. Female rats crossed the electrified grid when incentive males were in the incentive compartment significantly more frequently than when a receptive female was in the compartment.

In a modern neurochemical context, when gonadectomized females without hormone replacement were allowed a choice among four incentive animals, the females spent an equal amount of time in proximity to a sexually active male, a gonadectomized male, a female in proestrus, and a gonadectomized female. After estrogen priming, females exhibited a strong preference for proximity to the sexually active male, and this preference was further heightened by the addition of progesterone (Dudley and Moss, 1985). Emery and Moss (1984) demonstrated the importance of the VMH to motivational aspects of female sexual behavior. In an apparatus that allowed the females to pace contacts with a male, females bearing VMH lesions rarely permitted coital contact. However, when lesioned females permitted mounting, their probability of displaying lordotic behavior was similar to that of sham-operated animals (Emery and Moss, 1984).

Meyerson and Lindstrom (1971, 1973) used assays in which female rats were required to perform arbitrary responses or to cross obstructions to reach males. In their open-field tests, the amounts of time females spent in the vicinity of incentive males were significantly increased by injecting the animals with estradiol benzoate. The ability of estrogen treatment to increase approaches toward a sexually vigorous male peaked at 2 and 3 days after injection with estradiol benzoate and was a monotonic function of estrogen dose. Meyerson and Lindstrom (1971, 1973) also found that estradiol benzoate significantly increased the number of times OVX female rats would cross over an electrified grid to get to an incentive male. However, the same dose of estrogen did not affect grid crossings to get to an incentive female or to an empty incentive compartment. Larger and longer effects were seen with higher doses of estradiol benzoate. Finally, in runway-choice experiments, Meyerson and Lindstrom (1971, 1973) found that preference for an intact male over a castrated male by OVX female rats was significantly elevated by treatment with estradiol benzoate compared to vehicle control injections. Once again, the degree of preference in these runway experiments for an incentive male over an incentive female rat was a function of the dose of estradiol benzoate given the test female rat.

Thus, with three different behavioral approaches, Meyerson and Lindstrom's (1971, 1973) results indicated that estrogen injections of an OVX female rat would increase the frequencies of arbitrarily chosen responses resulting in contact with a vigorous male. Because the arbitrarily chosen responses in the three types of experiments were different from each other, the results were unlikely to be an artifact of one particular behavioral technique. In those studies in which different types of incentive animals were used, the most striking results involved the seeking of contact with the male rather than with other types of rats, suggesting that estrogens were heightening a tendency that was specifically sexual rather than generally social.

## D. Hypothalamic and Preoptic Mechanisms Involved in Two Types of Motivational Change

Estradiol heightens at least two forms of sex drive that have an interesting relationship to each other. They foster appetitive responses (in reproduction these comprise courtship behaviors), and then consummatory responses (in female mice, lordosis behavior).

*First,* by actions on the locus coerulens and other components of the ascending reticular activating system, estrogens and progestins combine to produce an aroused, muscularly taut animal. In particular, ascending neural projections to the POA combined with estrogen actions in the POA itself foster locomotion in a form that serves as an approach and courtship response (for details see chapter II). It has long been known that the administration of estradiol leads to a great increase in locomotor activity under circumstances in which locomotor wheels are essentially part of the female rat's home cage (Fig. 5) (Wang, 1923; Wade and Zucker, 1970; Fahrbach, Meisel, and Pfaff, 1985; Brown et al., 1993). Progesterone amplification of the estrogen effect, especially fosters peculiar forms of locomotion including hopping and darting, accompanied by tremendous tension in axial muscles, leading to head oscillations that give the appearance of "ear wiggling" (Whalen, 1974). Alternate approaches by the female to the male and subsequent withdrawals com-

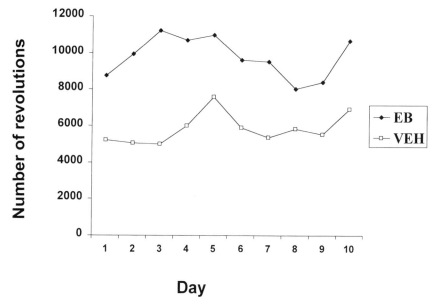

**Fig. 5.** Estradiol benzoate (EB) implanted in a silastic capsule subcutaneously in OVX Swiss Webster female mice led to higher rates of locomotor wheel activity than control subcutaneous implants with vehicle (VEH). (From Garey et al., unpublished data.)

prise female-paced mating behavior (Erskine, 1989; Brandling-Bennett, Blasberg, and Clark 1999).

Since solicitation behavior in estrous female rats follows a considerable increase in locomotion and itself includes a peculiar form of locomotion, and since estrogens followed by progestins—occurring naturally or administered experimentally—lead to both of these results (for reviews see Erskine [1989]; Fiber and Etgen [1997, 1998]; Ansonoff and Etgen [1998]), it is logical to think of a series of events in which the administration of hormone leads to locomotion that includes approaches to social targets and effectively comprises one aspect of courtship behaviors. How does this come about? The work of Professor Yasuo Sakuma and colleagues in Tokyo has analyzed regions of the POA important for courtship behaviors (Sakuma and Pfaff, 1980). Electrical stimulation of the medial preoptic area (MPOA) in awake, nonrestrained female rats stimulated locomotion but suppressed the lordosis reflex. In turn, estrogen facilitated preoptic neurons with outputs that would foster proceptivity (including locomotion in the form of hopping and darting) in the female rat, while estrogen simultaneously inhibited preoptic neurons with outputs that would suppress receptivity (Fig. 6). We infer that distinct subsets of estrogen-sensitive neurons in the POA of the rat differentially regulate proceptive and receptive behaviors (Kato and Sakuma, 2000). It is striking that Sakuma's conclusions based on work with rodents comport very well with the suggestion from Balthazart et al. (1998) that subregions of the preoptic medial nucleus differentially regulate appetitive and consummatory male sexual behaviors in Japanese quail. Likewise, Pfaus, Smith, and Coopersmith (1999) found appetitive "level-changing" behaviors statistically independent of consummatory measures in hormonally primed, sexually experienced female rats. All of these findings raise the possibility that in the vertebrate forebrain, as in certain insect nervous systems (Gammie and Truman, 1997), an individual compound such as a hormone or a neuropeptide can trigger the sequential performance of two behaviors. In the female rodent, as in the hawkmoth, *Manduca sexta,* a modulator can activate the first behavior, and then the first behavior is turned off while activating the second.

These preoptic neurons find their physiologic importance, in part, by their projections to the midbrain locomotor region (MLR) (Iwakiri et al., 1995). In the female rat's high state of tension, simultaneous activation of left and right MLRs, rather than normal left-right alternations, could change normal stepping into the hopping-and-darting characteristic of the proceptive female rat.

Controls over locomotion from lower in the neuraxis have been studied extensively (Kiehn et al., 1998). In particular, midbrain locomotor-related neurons are important for activating the reticulospinal controls of spinal rhythm–generating mechanisms (Rossignol, 1996). As well, rhythmic discharges within a cell group in the lower brain stem, the dorsal accessory olivary nucleus, are correlated with rhythmic limb stepping and, indeed, are enhanced by increases in circulating estrogens and progestins (Smith, 1997). Neurochemically, as well, preoptic neurons can be drawn into these courtship mechanisms, because estrogens have been demonstrated to increase protein kinase C (PKC) catalytic activity in the medial POA of OVX female rats, even under circumstances in which the hypothalamus is not affected (Ansonoff and Etgen, 1998). Figure 6 depicts these preoptic neurons (Turcotte and Blaustein, 1999) affecting locomotion through lower brain stem locomotor mechanisms. Note the important opposition between preoptic and medial hypothalamic cell groups relevant for reproduction.

Why would the female wish to make repeated and frequent contacts with the male? The obvious answer, positively Darwinian, is that such be-

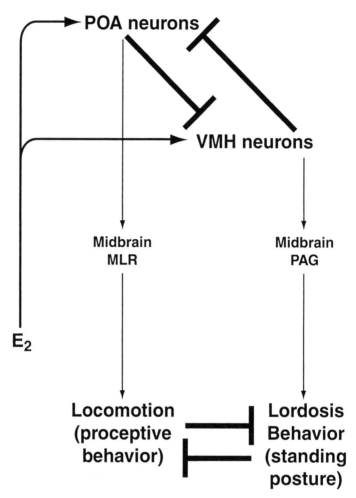

**Fig. 6.**   Estrogens ($E_2$) affect a subset of preoptic neurons that, operating in part through the MLR, foster locomotion and proceptive behaviors. Estrogens also operate through VMH neurons that, through their projections to the midbrain PAG, foster lordosis behavior. Thus, an individual hormone such as estradiol can turn on two separate types of motivational states important for reproduction. The inherent mutual neurophysiological opposition between POA neurons and VMH neurons enforces sequential performance of proceptive behaviors (involving locomotion) and lordosis behavior (involving a standing posture). Indeed, the female cannot locomote and exhibit lordosis behavior at the same time. The opposing neurophysiological forces between POA and VMH neurons permit an alternating temporal sequence of these two behaviors, locomotion and lordosis, which turns out to be biologically adaptive.

havior by the female will increase the probability of pregnancy and thus continued survival of the species. In turn, the pain control mechanisms discussed in chapter III encourage continued contact with the male despite the strong somatosensory stimuli involved.

A *second* manifestation of sexual motivation, meanwhile, emanates from actions of estrogen in the VMH that turn on several genes and elevate electrical activity in a specific subset of neurons (Pfaff, 1999a). This particular subset of neurons (Fig. 7) has extremely low spontaneous ac-

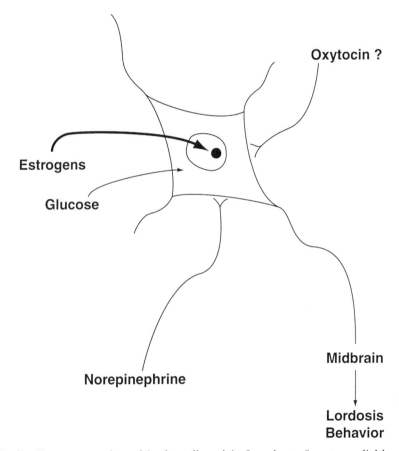

**Fig. 7.**   Estrogens are bound in the cell nuclei of a subset of ventromedial hypothalamic neurons that have very low spontaneous activity (Bueno and Pfaff, 1976). Neurons in this part of the VMH also can have their activity facilitated by glucose (see Fig. 8), norepinephrine (signaling arousal), and, possibly, oxytocin (Kow et al., 1991). Thus, these ventromedial hypothalamic neurons would be in an ideal position to integrate different neurophysiological and neurochemical forces fostering reproduction.

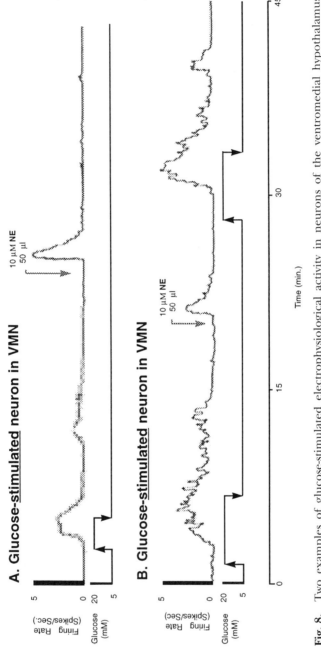

**Fig. 8.** Two examples of glucose-stimulated electrophysiological activity in neurons of the ventromedial hypothalamus (VMN). In addition, during these single-neuron electrophysiological recordings in hypothalamic tissue slices taken from female mice, norepinephrine (NE) was applied during low spontaneous activity in order to prove that the neuron was indeed there and sensitive to exogenous agents. (From Yang, Kow, Pfaff, and Mobbs, unpublished data.)

tivity, which then is elevated by estrogens (Bueno and Pfaff, 1976). This group of ventromedial hypothalamic cells (Fig. 8) also is interesting because it responds to glucose (signaling nutritional levels adequate for reproduction) and to norepinephrine (from the ascending reticular activating system). Therefore, such a set of hypothalamic neurons acts as a perfect integrator for the control of reproductive behavior, because they respond to a sex hormone, nutritional signals, and signals indicating an aroused state.

Estrogen also turns on the gene for the opioid peptide enkephalin in VMH neurons. Since ventromedial hypothalamic cells project to the ventrolateral PAG (vlPAG), an important site for analgesia, and since an opioid peptide itself should foster analgesia, it is clear that estrogenic actions through the VMH could contribute to an analgesic effect. In turn, through descending pathways that serve both reproductive behavior and analgesia, descending signals to the spinal cord permit lordosis behavior, instead of avoidance responses, to strong cutaneous and visceral stimuli applied by the male (for details see chapter III).

Many experiments indicate that preoptic physiological mechanisms often oppose hypothalamic mechanisms, and vice versa. Moreover, at the behavioral level, it is obvious that the female must stop locomoting in order to perform lordosis behavior, and vice versa. It appears that the natural opposition (Fig. 6) between the mechanisms of these two types of estrogen-dependent drives—locomotor courtship versus lordosis behavior—enforces the easily observed sequential relationship between them: first, courtship, then lordosis behavior, then courtship, then lordosis, and so forth.

Thus, we conclude that behavioral and neurophysiological evidence (Pfaff, 1980) indicate mutually inhibitory relations between the POA and the medial basal hypothalamus, both regions operating through the midbrain. The POA fosters locomotor behaviors as evidenced in female rodent courtship responses, as opposed to the medial basal hypothalamus, which enforces a standing response as required for lordosis behavior. It appears that the performance of courtship/locomotion "sets up" the initiation of the standing response coupled with lordosis. Likewise, the male's dismount releases the female from lordosis, leading to a new bout of locomotion. Between lordosis and locomotion, the termination of each rebounds to the other.

The underlying concept is that, as opposed to a static view of a set of individual instinctive behaviors, the very opposition between two behavioral responses—it is impossible to locomote and stand still at the same time—enforces an alternating time series between these two responses. Elaborating on this point, the neurophysiology supports a continuum of response mechanisms: basic locomotion, as required for approach behaviors by the female, in turn comprise parts of courtship behaviors, which in turn are interpreted as evidence of motivation, or sex drive. The additional element in the temporal series of responses exhibited by female rodents is that the halting of locomotion allows the male to mount, causing lordosis behavior, whose cessation in turn springs a new bout of locomotion, leading to another lordosis response, and so on.

## E.  Summary

With the intention of elucidating the simplest, deepest, broadest forms of motivation necessary to explain changes in behavior, many theorists have explored the concept of central motivational states. Such concepts might apply to complex units of mating behaviors that are not limited to hypothalamically controlled lordosis, but would also include locomotor, exploratory, and seeking behaviors dependent on the POA and midbrain. Mutual opposition between preoptic mechanisms (for locomotion) and hypothalamic mechanisms (for lordosis behavior) generates a temporally alternating series between the two behaviors.

For these events to occur, at least two broad fluctuations in neural states are important to consider. In chapter II, we treat ascending arousal systems that would be required for the exploratory behaviors we have already reviewed and that would be activated by sexually relevant and painful stimuli. Then, in chapter III, we cover opioid mechanisms of analgesia that are invoked following sexual or painful stimulation.

# II

## *Ascending Arousal Systems Activated*

We all know the feeling. While preparing for an important meeting at work and, at the same time, managing a fractious kid at home, we have lost the keys to the house and the car. Upset about that, we have spilled the stuff that was almost ready for the lunch bags. Things are coming apart. On a continuum from comatose through pleasantly stimulated through frazzled (Arnsten, 1998), we are over the top. Are we being rattled all too easily? Who knows? Well, the biology of human excitability and temperament, including understanding its genetic dispositions, now represents an extremely active field of research (Svrakic, Przybeck, and Cloninger, 1992, Svrakic et al., 1993; Stallings and Hewitt, 1996). In general terms, the physiological bases of arousal in its optimum manifestations—in its failure (stupor) and in its excesses (nervous disorganization)—are ready for scientific attack.

Dealing with the neurophysiology, neurochemistry, and molecular biology of ascending arousal systems is especially important for this book. The broad arousal states influenced by sexual and painful stimuli are first realized in these ascending systems. It is precisely the broad scope of neural systems serving arousal that allows sexual and painful stimuli, in their normal and abnormal aspects, to overlap.

The earliest electrophysiological results on ascending arousal systems and their subsequent conceptualization led to the notion of a monolithic brain stem mechanism whose activation would excite large parts of the forebrain, with an emphasis on thalamic and cortical neuronal regions. For example, the writings of Horace Magoun and Donald Lindsley, which we review subsequently, gave rise to this impression. Certainly, the most widespread manifestations of neuronal systems ascending from the midbrain and hindbrain could include the activation of behavioral responses relevant to sex and pain and could be fundamental to the most primitive aspects of human consciousness.

The separate histochemical systems containing norepinephrine, dopamine, and other monoamines, initially discovered and elucidated by neurobiologists at the Karolinska Institute, now have been distinguished and further fractionated functionally and anatomically. These are reviewed in sections A and B. Moreover, at the behavioral level (Robbins and Everitt, 1996; Posner, 1995) the concept of arousal has needed to be subdivided. Those aspects relevant to sociability and reproduction (section C) and to human awareness (section D) are of greatest interest here. These systems, broadly tuned as they are, represent important components of the overlaps between sex and pain.

## A.  Ascending Reticular Activating Systems

According to the writings of Magoun (1958), interest in the deep-lying midline structures of the brain stem goes back to the eighteenth century, when neuroanatomists such as Thomas Willis suggested their central role in mental activities. Subsequently, after many decades of concentrated physiological research on the reflex functions of the spinal cord and on the higher mental activities dependent on the cerebral cortex, electrophysiological and neuroanatomical work once again returned to the central roles played by brain stem mechanisms, which Magoun (1958) called "non-specific reticular systems." Magoun's interests, and those of his collaborator, Moruzzi (Moruzzi and Magoun, 1949), focused on electroencephalogram (EEG) recordings from the cerebral cortex. Most striking was their demonstration that direct electrical stimulation of the brain stem reticular core reproduced all the electrocortical features observed in the EEG arousal reaction associated with natural wakefulness: electrocortical recordings distinguished by a change from low-frequency, high-amplitude activity (typical of a quiet, nonresponsive animal) to high-frequency, relatively low-amplitude activity (typical of an alert, responsive animal). Conversely, experimental destruction of the central brain stem left an animal that appeared deeply asleep or anesthetized following the operation (French, 1952). The importance of the medial placement of these brain stem regions was obvious by contrasting their results with the sensory neglect (without impairment of wakefulness or EEG activity) following laterally placed midbrain lesions (Sprague, Chambers, and Stellar, 1961). While the great majority of articles deriving from this early electrophysiological work concentrated on the fundamental properties of behavioral wakefulness and EEG arousal, even then, some laboratories brought in a connection to neuroendocrinology. Sawyer and Kawakami (1961) noted that the enhanced alertness and augmented motor performance during the appetitive and exploratory stages of mating behaviors were associated with an increased excitability of the reticular system as measured by heightened EEG arousal. In the face of an increasing tendency to subdivide these brain stem reticular systems neurochemically, electrophysiologically, and functionally, Sherin et al. (1996) suggested that certain components of these ascending systems are coalesced in their impact on neurons of the lateral POA associated with sleep and wakefulness.

Experiments conducted by Donald Lindsley and associates included many behavioral observations as well as changes in the cortical EEG. Placement of lesions of the midbrain tegmentum of cats in positions to interrupt fibers of the ascending reticular activating system resulted in chronic somnolence (Lindsley et al., 1950). These results were distinguished from those of lesions of specific sensory pathways or lesions confined to the PAG, which did not lead to such somnolence. For example, in a cat with damage to the midbrain midline tegmentum at a rostral level, just behind the diencephalon, postoperative results showed that it would lie motionless on its side with its eyes closed, as though deeply asleep, for 21 days (Lindsley et al., 1950). EEG changes after such lesions revealed that the EEG activation pattern of high-frequency, low-amplitude waves was reduced or abolished (Lindsley, Bowden, and Magoun, 1949). These massive deficits following midbrain damage could be distinguished from results following damage of the medulla, not associated with such EEG changes. A full decade of work in many laboratories, following these seminal articles (for a review see Lindsley [1960]), confirmed that damage to the midbrain tegmentum was associated with somnolence or sleep with no sign of behavioral arousal, whereas high-frequency electrical stimulation of the brain stem reticular formation would produce characteristic arousal and alerting responses.

Tomas Hokfelt, Kjell Fuxe, Annica Dahlstrom, and colleagues at the Karolinska Institute (Dahlstrom and Fuxe, 1964; Fuxe, 1965; Hamberger and Hokfelt, 1968), in Stockholm, Sweden, carried out the first corresponding neurochemical and neuroanatomical work. Quickly, their results revealed fractionations and subdivisions of ascending brain stem systems that had widespread projections in the forebrain (see review in Fuxe, Hokfelt, and Ungerstedt, [1970]). Thus, the cells of origin of ascending noradrenergic systems and serotonergic systems were identified in the hindbrain and the midbrain, respectively, and some of their very broad fields of innervation were described in the limbic forebrain and the neocortex. A tuberoinfundibular dopamine system of neuroendocrine importance was discovered (Hokfelt, 1967) and dopaminergic nerve terminals in the limbic cortex, derived from cell bodies in the brain stem, revealed by a combination of pharmacological and histochemical techniques (Hokfelt et al., 1974). Furthermore, ultrastructural investigations of the granular vesicles in monoamine neurons laid the morphological basis for physiological studies of the release of noradrenaline, dopamine, and serotonin (Hokfelt, 1967, 1968, 1969). Thus, even though these ascending pathways crucial for "waking up" the forebrain could be distinguished in their widespread fields of termination from specific sensory pathways with localized punctate connectivities, the work just quoted demonstrated that they also could be chemically distinguished from each other, leading to the suggestion of functional distinctions as well.

Roles for ascending *noradrenergic* pathways in the production of alertness, arousal, and adaptive behavioral responses are well accepted. In support of behavioral arousal, autonomic controls connected with noradrenergic cells of the lower brain stem have also been laid bare. Projections from the nucleus of the solitary tract, the "A2" group of noradrenergic neurons, to the rat amygdala have been connected with controls by the amygdala over cardiovascular and other visceral functions (Zardetto-Smith and Gray, 1990). In addition, axonal projections from catecholaminergic neurons within the ventrolateral medulla (medullary "A1" adrenergic neurons) to limbic structures have been identified and are assumed to be involved in relaying afferent information directly to the amygdala, especially its central nucleus, which is thought to be involved in the integration of autonomic, endocrine, and behavioral functions (Roder and Ciriello, 1993, 1994). By contrast, *dopaminergic* neurons projecting from the midbrain to the limbic forebrain have been intimately connected with the initiation and adaptation of directed motor responses (Le Moal and Simon, 1991). Put most succinctly, these neurons appear to be required for directed movements toward stimuli that have a valence of positive or negative reward. Frequently, these are termed *appetitive behaviors* (Blackburn, Pfaus, and Phillips, 1992).

Now, it seems, the far-reaching concepts intrinsic to ascending reticular activating systems have been subject to intense examination (Steriade, 1996) and new experimental work. For example, activation of the mesencephalic reticular formation has recently been claimed to enhance stimulus-specific synchronization of neuronal spike responses in the visual cortex (Munk et al., 1996). The import of these systems for neuroendocrine functions has not been neglected. Catecholaminergic inputs to hypothalamic nuclei such as the paraventricular and supraoptic nuclei have been charted in detail (Swanson et al., 1981; Swanson and Sawchenko, 1983). Combinations of anterograde transport, retrograde transport, and immunohistochemical techniques suggested that, while adrenergic inputs to these hypothalamic nuclei arise from distinct medullary cell groups and thus might relay different types of sensory information, these inputs can be integrated by similar groups of hypothalamic neurons that regulate neurosecretory and autonomic functions (Cunningham, Bohn, and Sawchenko, 1990). It is clear, then, that noxious or other salient stimuli activate medullary catecholamine neurons (Jones and Blair, 1995) that can process and relay this sensory information to hypothalamic neurons with neuroendocrine roles (Smith and Day, 1994).

## B. Structure of Arousal States

Attempting to analyze CNS arousal states heightened by both sexual and painful stimuli, we stumble onto fundamental questions about the behavioral, statistical, neurophysiological, and neurochemical components of these states (Fig. 9).

Some neurophysiologists and behaviorists have emphasized the unity of certain CNS arousal mechanisms. Others have argued for multiplicity. For an example of the latter, in a rhetorical attack on some of the tenets of arousal theory, Eysenck (1982) attempted to damage the notion of "generalized arousal." He argued against the Yerkes-Dodson law, which posits an inverted-U relationship between arousal and the performance of learned behavior. This law states that there must be "an optimum level of arousal" below which and above which the performance of learned responses would decline. Eysenck (1982) argued too strongly against that field of work. Obviously, a person in a state of reduced arousal, as in a vegetative state, could not perform well. Equally clear is the fact that a person in a state of blind panic, disastrously heightened arousal, could not do his or her best. Therefore, in the broadest terms, Yerkes-Dodson must be correct; there must be a peak of performance in between.

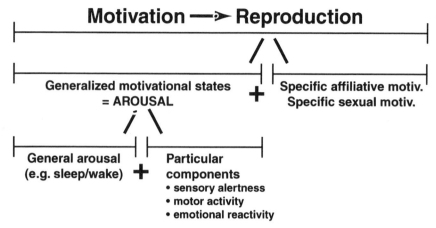

**Fig. 9.** Hormonal activation of reproductive behavior reflects hormonal alteration of an underlying motivational state. During the past few decades, experiments and theories on motivation have shown that individual states are composed of both specific and generalized components. The generalized components often go under the name "activation" or "arousal." In turn, arousal states are composed of both generalized (e.g., the sleep/wake cycle) and particular components, frequently involving aminergic ascending brain stem systems. Genetic and hormonal influences on these ascending systems will provide for a fascinating series of new experiments.

Likewise, Robbins and Everitt (1996) argued against a straw man, "a unitary concept of arousal." They constructed false dichotomies ("perfect unity" versus "perfect uselessness") and then proclaimed "doubts in the utility of arousal-like constructs"(703). In addition, their experiments were confusing because they addressed such complex, multicomponent tasks for which clear interpretations were quite impossible.

Systematic rigorous analysis of behavioral results is much to be preferred (Frohlich et al., 1999), especially in genetically manipulable organisms such as mice. In a study with a large body of behavioral data from female mice, analyzed by factor analysis and cluster analysis, Frohlich (Frohlich et al., 2001) found that a one-factor solution accounted for 29 percent of the variance (Fig. 10). Thus, a generalized arousal exists. However other, particular arousal components do as well (see below) (Fig. 11).

Thus, the situation is much more interesting neuroanatomically, neurochemically, and functionally than the scenario some theorists whipped up when they imagined a false dichotomy between "unitary arousal" and "no arousal." Indeed, not all autonomic measures of arousal correlate with each other (Thayer, 1970), but demonstrably they can correlate well with verbal self-reports of arousal state. Eysenck (1982) apparently was arguing to reduce the influence of the reticular formation–based thinking that stemmed from the work of Moruzzi and Magoun, quoted in section A and summarized in the popular book *The Waking Brain* (Magoun, 1958). Even as he argued, physiologists and anatomists studying the reticular formation had begun to subdivide its functions according to cell group, projection, and neurochemistry (for a review see Hobson and Brazier [1980]). Obviously, arousal has both generalized and particular components, as does its supraordinate concept, motivation. The neuroanatomical and neurochemical bases of this dual character of arousal functions are easily tabulated (Table 1). Functionally, arousal could be manifest, at least, in heightened sensory alertness and attention, motor activity under a variety of conditions, and emotional and autonomic reactivity (Frohlich et al., 1999). In the aforementioned experiments with mice, using a variety of assays related to arousal, stable factors and clusters of arousal components emerged (Fig. 11). Only some of these were intuitively obvious (Frohlich et al., 2001, in press).

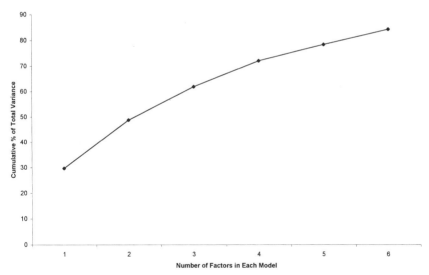

**Fig. 10.** A large number of OVX female mice were subjected to a set of assays measuring sensory, motor, and emotional components of arousal. The resulting cross-mouse correlation table was subjected to factor analysis. The single-factor model accounted for 29 percent of the variance, indicating that a "general arousal state" does exist but accounts for a minority of the variance. More than 80 percent of the variance was accounted for by the time six factors were included. (From Frohlich et al., 2001.)

**6 Factor Model**

| Factor # | Variables | Loadings |
|---|---|---|
| 1 | Total Distance(OFT) | .971 |
|  | Center Distance (OFT) | .778 |
|  | Margin Distance (OFT) | .968 |
| 2 | Total Nosepokes | .957 |
| 3 | Running Wheel | .786 |
|  | Center Time (OFT) | .808 |
| 4 | Acoustic Vmax1 | .926 |
|  | Acoustic Vmax2 | .928 |
|  | Acoustic Vavg1 | .948 |
|  | Acoustic Vavg2 | .927 |
| 5 | **Tactile Vmax** | .903 |
|  | **Tactile Vavg** | .929 |
| 6 | Nosepoke Duration | .363 |
|  | **Freezing to Context** | .636 |
|  | **Freezing to Tone** | .873 |

**6 Cluster Model**

| Cluster # | Variables |
|---|---|
| 1 | Total Distance(OFT) |
|  | Center Distance (OFT) |
|  | Margin Distance (OFT) |
| 2 | Total Nosepokes |
| 3 | Running Wheel |
|  | Center Time (OFT) |
| 4 | Acoustic Vmax1 |
|  | Acoustic Vmax2 |
|  | Acoustic Vavg1 |
|  | Acoustic Vavg2 |
|  | **Tactile Vmax** |
|  | **Tactile Vavg** |
| 5 | Nosepoke Duration |
| 6 | **Freezing to Context** |
|  | **Freezing to Tone** |

**Fig. 11.** There is good agreement between the six-factor analysis model and the six-cluster analysis model of the data emanating from the arousal-based experiment on OVX female mice described in Fig. 10. Measures presumed to reflect sensory, motor, or emotional components of arousal were derived from open-field tests (OFT), a home cage exploratory measure (nose pokes), locomotor wheels, acoustic startle assays, tactile startle assays, and conditioned freezing as part of fear conditioning. (From Frohlich et al., 2001.)

Interestingly, experimental approaches to normal arousal in humans and in the laboratory animals we use, such as mice, map neatly onto the "dimensions of anesthesia." In humans, the signs of anesthetic depth during surgery are related to, but separate from, the clinical considerations in chapter III, section G. Derived from a long history of clinical experience, authoritatively reviewed (Miller, 1994; Nimmo, 1994), and further rationalized by the use of quantitative electrophysiology (Glass et al., 1997), these dimensions of anesthesia that allow the practitioner to gauge depth are as follows: (1) awareness of sensory stimuli; (2) analgesia; (3) muscular, motoric activity; (4) autonomic responsiveness (as indexed, e.g., by heart rate variability); and (5) memory for events during surgery. It probably is no accident that the first two criteria are closely

**Table 1. Ascending Arousal Systems Have Both General
and Specific Components**

General features of arousal systems
  1. Anatomical connections are not point-to-point.
  2. Arousal effects are not limited to individual stimulus/response combinations. Instead, they are prerequisites for the occurrence of particular stimulus/response combinations.
  3. There is considerable redundancy, neurochemically and functionally, across arousal components.

Particular features of arousal systems
  1. Cell bodies of NE, DA, 5-HT, ACh, and His systems are different, so inputs are not identical.
  2. Axonal distributions are different, so target regions are not identical.
  3. The greatest functional impacts are not identical among NE, DA, 5-HT, ACh, and His.

NE, norepinephrine; DA, dopamine; 5-HT, serotonin; ACh, acetycholine; His, histamine.

related to the major part of this book, chapters II and III. In fact, the formulation of reproductive behavior mechanisms in chapter IV depends heavily on hormonal effects on arousal and analgesia.

All of the arousal subsystems contribute to the hormone-dependent CNS states treated in this book and are being confronted in their neuroanatomical and neurochemical details. Perhaps the simplest summary of the situation as can be envisioned now is depicted in Fig. 12. Plunging right in, we start with the most difficult, least clear subsystems.

## 1. Cholinergic, Serotonergic, and Histaminergic Mechanisms

Both cholinergic and serotonergic systems have been conceived as contributing to electroencephalographic activation. Vanderwolf (1988) made a case for cholinergic and serotonergic contributions based on combinations of pharmacological and electrophysiological evidence. In addition, he relied on his own evidence (Vanderwolf and Baker, 1996), using drugs such as reserpine that would block monoaminergic systems, as well as lesion evidence from other laboratories (Jones, Harper, and Halaris, 1977; Crow et al., 1978), in order to say that ascending noradrenergic and dopaminergic systems are not required for electroencephalographic activation.

The likely importance of ascending serotonergic systems is evident immediately on visualization of the widespread serotonergic innovation of

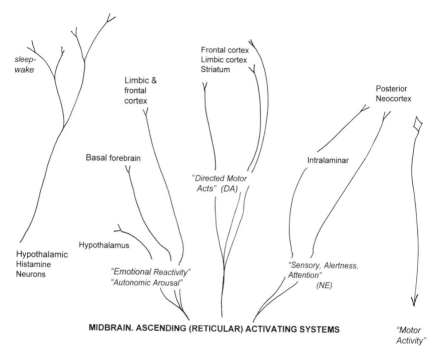

**Fig. 12.** Ascending reticular activating systems rising from the hindbrain and midbrain and traveling through the midbrain to influence electrical activity and behavior through forebrain neurons have been subdivided neuroanatomically, neurochemically, and functionally. Substantial interruption of these systems by mechanical or vascular accidents damaging the midbrain or the midline thalamus, particularly if ascending nonspecific cholinergic systems are also interrupted, can lead to patients in coma or vegetative states. Further functional analyses in conscious human beings with or without attention deficit disorder are under investigation (Swanson, Fossella, Pfaff, and Posner, in progress) and are being backed up by functional genomic studies in mice (Fossella et al., unpublished data). DA, dopamine; NE, norepinephrine.

limbic and forebrain structures (Conrad, Leonard, and Pfaff, 1974; Moore, Hataris, and Jones, 1978). Not only direct fiber projections to cortical regions but also indirect actions of serotonin, affecting cholinergic neurons in the basal forebrain, are likely to be important for cerebral activation (Jones and Cuello, 1989; Cape and Jones, 1998). Both ascending serotonergic and noradrenergic systems may act in part via cholinergic nucleus basalis neurons (Fort et al., 1995). Serotonergic agonists would activate the EEG in the sense of suppressing low-frequency activity (Dringenberg and Vanderwolf, 1996), while antagonists at serotonin receptors would clearly block atropine-resistant cerebral activation

(Vanderwolf, 1984; Vanderwolf et al., 1989; Watson et al., 1992; Dringenberg and Vanderwolf, 1997). Recent evidence from gene knockout mice suggests that $5\text{-HT}_{1B}$ receptors might be especially important (Boutrel et al., 1999).

The role of cholinergic neurons in serving monoaminergic effects on cerebral electroencephalographic activation is exemplified by the ability of a cholinergic antagonist, scopolamine, to abolish the stimulation effects from the locus coeruleus (LC) in the activation of the EEG (Dringenberg and Vanderwolf, 1998). Widespread ascending cholinergic projections presumably provide the anatomical basis for these important cholinergic contributions to encephalographic activation (Jones, 1993; Mesulam, 1995). Neurochemical changes in cholinergic neurons as revealed by c-fos immunocytochemistry (a histochemical technique for demonstrating neuronal genetic activation) also drew in these neurons to changes during sleep (Maloney, Mainville, and Jones, 1998). More important, cholinergic blockers could reduce low-voltage, high-frequency activity in the neocortex of waking rats (Stewart, MacFaber, and Vanderwolf 1984; Vanderwolf and Stewart, 1986). It may not be just an accident that the same mesopontine cholinergic system that modulates arousal is also responsible for acetylcholine-mediated analgesia (Iwamoto, 1991; Iwamoto and Marion 1993a, 1993b). Both strong sexual and painful stimuli should activate this system. Through these neurons, therefore, links between sex and pain are constructed in both their contribution to arousal and analgesia.

Histamine is clearly associated with the sleep/wake cycle, explaining, for example, why antihistamine agents used for allergies make people sleepy. A cellular basis for this is revealed by the ability of histamine to electrophysiologically excite cholinergic neurons in the nucleus basalis (Khateb et al., 1995). While involvement of H-3 receptors was excluded on a pharmacological basis, both H-2 and H-1 receptors were required for a full activation of cholinergic neurons (Khateb et al., 1995). Nevertheless, the presence of low-voltage fast activity in the neocortex of animals whose neocortical histamine had been depleted suggested to Vanderwolf and colleagues that histamine does not play an absolutely essential role in electroencephalographic activation during waking (Servos et al., 1994).

Serotonergic links to hormonally driven functions further emphasize the importance of this ascending system with an impact on the emotional, hormonal, and autonomic aspects of arousal. Studies have proved that serotonergic $5\text{-HT}_{2A}$ receptors are important for the estrogen-induced surge of luteinizing hormone–releasing hormone (LHRH) (Fink et al., 1999), and specific serotonergic subtypes have been drawn into the explanation of controls over female rat lordosis behavior (Uphouse et al., 1996a, 1996b; Maswood, Caldarola-Pastuszka, and Uphouse, 1997; Trevino et al., 1999). The researchers conducting these studies may well have been documenting how serotonergic contributions to arousal states could influence coordinated release of luteinizing hormone (LH) and lordosis behavior. Furthermore, the effects of synaptic serotonin levels on mood have been made evident (see Section E).

In summary, widespread serotonergic projections to limbic forebrain, neocortex, and, importantly, basal forebrain cholinergic neurons support EEG arousal and are also likely to be involved in the emotional and autonomic changes associated with impulsive behaviors and more generally with various mood states. The coordinated contributions of cholinergic neurons have been revealed, in part, by cholinergic blocking agents but also by the necessity of basal forebrain cholinergic excitation to mediate the effects of ascending serotonergic and noradrenergic systems. Besides those nucleus basalis cholinergic neurons, important cholinergic cell groups reside in the lateral dorsal nucleus of the tegmentum and in the peripeduncular nucleus. In turn, these work at least in part through thalamic relays.

Mesopontine cholinergic neurons and perhaps glutamatergic cells of the brain stem reticular formation activate thalamic neurons that project to the cortex during states of waking and rapid eye movement (REM) sleep (Steriade, Contreras, and Amzica, 1997). Both short-lasting nicotinic depolarization associated with increased membrane conductance in thalamocortical neurons and long-lasting muscarinic depolarizations associated with increased input resistance are involved. The anatomical basis of these physiological phenomena is found in a strong array of cholinergic synapses whose cell bodies of origin are in the lateral dorsal tegmental nucleus and the peripeduncular area, in addition to noradrenergic inputs that originated in the LC (see Steriade, Contreras, and Amzica, [1997]). As a result, thalamocortical systems with diffuse projections in the telencephalon are under the control of ascending systems spread throughout the brain stem core. The prominent participation of mesopontine cholinergic neurons, emanating from their intrinsic electrophysiological properties, is shown by their state-dependent activity: increased firing rates and enhanced excitability in advance of behavioral wakefulness (Steriade, Contreras, and Amzica, 1997).

## 2. *Noradrenergic Mechanisms*

Long and widespread projections from a norepinephrine-producing cell group, the LC, innervate the intralaminar and midline nuclei of the thalamus, striatum, basal forebrain, and cerebral cortex (Jones, 1991a; Jones and Yang, 1985). This type of observation with tritiated amino acid autoradiography confirms and extends the fluorescent histochemical work of Fuxe, Hokfelt, Ungerstedt, and Dahlstrom. Together with neighboring cholinergic neurons of the brain stem at upper pontine and lower mesencephalic levels, noradrenergic LC neurons yield pharmacological, lesion, and electrophysiological data that qualify them for important roles in behavioral attention and arousal (see review in Jones [1991a, 1991b]). Not all of the influences of LC cells on the cerebral cortex are necessarily direct. An important indirect route of influence would include the effects of noradrenergic LC projections to cholinergic cells in the basal forebrain (Jones, 1991b; Cape and Jones, 1998).

A well-analyzed cognitive function is that of sustained attention (Posner and Petersen, 1990). From pharmacological behavioral studies, norepinephrine has been inferred to facilitate attention especially for prolonged periods or during stress (Marrocco and Davidson, 1996). Thus, the experiments and thinking of Michael Posner and colleagues would indicate that norepinephrine is involved in maintaining an alert state propadeutic to higher cognitive processes. For example, noradrenergic antagonists administered to alert monkeys would eliminate the normal effect of warning signals in reducing reaction times (Posner, 1995). In their review (Robbins et al., 1998), Robbins and Everitt invoke the impaired accuracy in a five-choice serial reaction task—impaired due to lesions of the dorsal noradrenergic bundle—to implicate the coeruleur-cortical noradrenergic system in selective attention when tested under difficult circumstances. In neuroanatomical terms, noradrenergic ascending reticular systems eventually affecting parietal cortical regions would carry out such effects on sensory processes (Posner and Petersen, 1990; Coull, 1998). Such difficult circumstances include performance under stress, perhaps involving the ability of corticotropin-releasing factor to target LC dendrites (Van Bockstaele, Colago, and Valentino, 1998). For example, during LC electrophysiological recordings, electrical activity is high during waking and alertness, and slow during sleep (Aston-Jones et al., 1996). Reactions to imposed sensory inputs during higher LC activity reveal markedly better responsiveness to meaningful stimuli, those that have been designated as "target cues" in a behavioral task. Aston-

Jones et al. (1996) arrive at the conclusion that LC neurons are more than simple components of the sympathetic autonomic system and, instead, are part of an "urgent response system" of behavioral import. Bilateral 6-hydroxydopamine lesions of the LC that produced an 80 percent depletion of noradrenaline in the prefrontal cortex (but that avoided collateral damage to other ascending noradrenergic pathways) significantly reduced peak adrenocorticotropic hormone (ACTH) and corticosterone responses to acute restraint stress even under circumstances in which other physiological responses to a 4-wk period of chronic stress were not affected (Ziegler, Cass, and Herman, 1999). In monkeys, changes in either spontaneous or stimulus-induced patterns of LC neuronal electrical activity were, as expected, correlated closely with fluctuations in behavioral performance (Usher et al., 1999).

Finally, LC neurons participate in POA-related functions, some of which were referred to in chapter I. That is, even as preoptic neurons cooperate with mesencephalic locomotor region mechanisms in producing signs of arousal, locomotor activity, and courtship (see Sakuma references in chapter I), other portions of the POA are related to sleep (Osaka and Matsumura, 1994). Electrical stimulation of the LC by single pulses inhibited a majority of preoptic sleep-related neurons and excited a majority of waking-related neurons (Osaka and Matsumura, 1994). For the central theme of this book, it is also important that LC neurons, as well as A5 and A7 noradrenergic neurons, are involved in analgesic processes through interactions with the rostral ventral medulla and the dorsal horn of the spinal cord (see chapter III). In sum, noradrenergic LC outputs are involved in broadly defined behavioral functions connected to arousal, as well as in autonomic physiology, with an emphasis on sustained attention, especially during stress.

### 3. Dopaminergic Mechanisms

Two major ascending systems carry dopaminergic influences to the forebrain: the nigrostriatal system and the mesolimbic dopamine system. From a neuropharmacological point of view, it is seen that dorsal aspects of these projections, to the striatum, are centrally involved in sensory-motor integration, while projections to the basal forebrain are, in the words of Iversen (Dourish and Iversen, 1989; Martin-Iverson and Iversen, 1989; Iversen, 1998), "integral to motivational arousal."

In such experiments, for example, lesions of the nucleus accumbens (Nacc), centrally involved in reward, abolished the ability of amphetamines to increase locomotion. Another important projection, however, would be dopaminergic projections to the frontal cortex, involved in the direction of motor activities (as reviewed in Coull [1998]). Different fields of projection involve different dopamine receptors, and there are circumstances in which D1 and D2 receptors, for example, have opposing roles (Vanderschuren et al., 1999). Genetic knockouts to delete the dopamine transporter provide one of the most dramatic manipulations to prolong the lifetime of extracellular dopamine (Giros et al., 1996). Such genetically modified animals have marked spontaneous locomotion, as elevated as animals on high doses of psychostimulants (Giros et al., 1996; Jones et al., 1998).

A traditional point of view stated that dopaminergic projections to cell groups such as the Nacc would be intimately associated with the process of reward, because the electrical activity of certain dopamine neurons seemed to signal predictions of future rewarding events (Schultz, Dayan, and Montague, 1997). Dopaminergic fluctuations in the Nacc during rewarded acts support this point of view (Ranaldi et al., 1999). However, a broader perspective also seems appropriate. Dopaminergic projections to the basal forebrain seem most cogently to be involved in the selection of directed motor acts toward attended, salient stimuli. Even during experiments in which general motoric mechanisms are well controlled for, the interpretation that dopaminergic neurons support responses to "behaviorally significant stimuli" seems to be justified. For example, Professor Jon Horvitz at Columbia University discovered responses of dopaminergic neurons in the ventral tegmental area (VTA) to loud clicks or bright flashes of light (Fig. 13), which suggested that they would respond to salient stimuli regardless of whether those stimuli had acquired reward properties, and that physical characteristics such as sudden onset of the stimulus would be important (Horvitz, 1997). Conversely, lesioning of mesolimbic and nigrostriatal dopaminergic systems markedly slows latency to respond to salient stimuli, reduces the number of premature responses, and increases total omissions of response (reviewed in Robbins et al. [1998]). These functions might be particularly important during stress, because the imposition of acute stress led to a significant increase in the dopamine precursor, Dopa, in the medial prefrontal cortex (but not Nacc) (Nakahara and Nakamura, 1999).

These dopamine-related phenomena, relating back to the subject matter of chapter I, might be important for appetitive aspects of mating behaviors in both males and females. For example, extracellular dopamine levels in the medial POA are closely related to copulatory behav-

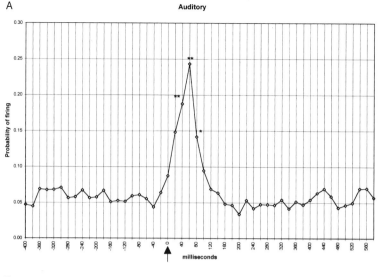

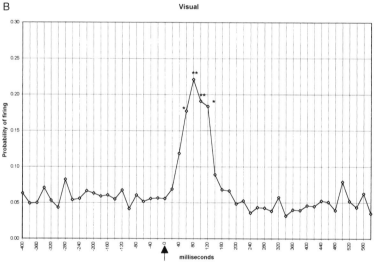

**Fig. 13.** The mean probability of ventral tegmental dopaminergic neuronal firing before and after the delivery of an auditory (click) and visual (light flash) stimulus (arrow) in freely moving cats (Horvitz, 1997) is shown. Probability estimates, at each time bin, were averaged across all dopaminergic neurons tested for an auditory (*A*) and visual (*B*) response. Asterisks mark the poststimulus time bins for which neuronal discharge probability was significantly greater than that observed during the prestimulus baseline (*$p < .05$, **$p < .01$). As shown here, mesolimbocortical dopaminergic neurons respond to salient (high-intensity, rapid-onset) sensory events across sensory modalities. These dopaminergic neurons respond to not only stimuli with salient physical characteristics, but also stimuli whose salience derives from their association with rewarding (Schultz et al., 1998) or aversive (Guarraci and Kapp, 1999) events. Similar single-unit responses are observed in dopaminergic neurons within the substantia nigra (Steinfels et al., 1983; Ljungberg, Apicella, and Schultz, 1991; Horvitz, 2000).

iors in male rats (Hull, Lorrain, and Mataszewich, 1997; Ellis and Ebertz, 1998). In females, the effects of progesterone metabolites on reproductive behaviors mediated through VTA neurons (Pleim, Lisciotto, and De-Bold, 1990; Frye and DeBold, 1993; Frye and Leadbetter, 1994; Frye and Gardiner, 1996) are highly likely to involve dopaminergic mechanisms. In sum, the ability of ascending dopaminergic influences to enhance animals' responses to salient stimuli of behavioral significance is likely to be important to the reactions of males and females to reproductively relevant signals including those important for sexual motivation, as referred to in chapter I.

All the neurochemical systems treated thus far–cholinergic, histaminergic, noradrenergic, and dopaminergic–are involved in their own ways in the processes of brain arousal. In what ways do their salient characteristics apply to the generalized motivational states contributing to reproductive behavior?

## C. Application to Sexual Behavior

Every aspect of the behavior of the estrogen-treated female rat indicates an aroused animal. There is a great increase in locomotor activity, and in preparation for courtship behavior, the estrogen and progesterone-treated rat displays a muscularly taut posture, in the extreme. Activation of these arousal systems and of locomotor-controlling neurons in the POA leads to a form of locomotion that, when suddenly halted (indicating that the animal is ready for lordosis), encourages the male to encounter the female in the proper position for reproduction (Pfaff, 1980).

### 1. Noradrenergic Systems

Estrogens address noradrenergic systems in a manner that might be important not only for the behaviors already discussed, but also for control of the release of LH release from the pituitary. Binding of radioactive estrogens by LC neurons opened the possibility that estrogens could directly address noradrenergic ascending systems (Pfaff and Keiner, 1973). Extended by Heritage et al. (1980), these binding studies left open the question, Is the classical estrogen receptor-α (ER-α) or the novel estrogen receptor-β (ER-β) active in brain stem noradrenergic neurons? In situ hybridization with radioactively labeled probes specific for the classical ER-α or the novel ER-β messenger RNA (mRNA) revealed, without a doubt, that LC (A6) and NTS (A2), both noradrenergic cell groups, contain mRNA for the classical ER (Shughrue, Komm, and Merchenthaler, 1996). Since these nerve cell groups also contain mRNA for ER-β, these results raised the possibility either that different neurons within these cell groups used different types of the estrogen receptor or that the well-known ability of ER-α or ER-β to heterodimerize might be functionally important in these neurons. In either case, it is clear that estrogen addresses noradrenergic nerve cell groups in the brain stem critical for arousal and alertness of the animal.

In absolute accord with those results were the findings of Simonian and Herbison in sheep (Simonian et al., 1998) and rats (Simonian and Herbison, 1997) that double-labeled immunocytochemistry revealed ER-α immunoreactive neurons and dopamine β-hydroxylase were discovered among brain stem A1 and A2 noradrenergic neurons. The use of a retrograde tracer, flurogold, injected into the POA of ewes, coupled with ER immunocytochemistry and dopamine β-hydroxylase immunocytochemistry indicated that the most caudal noradrenergic cells in the brain stem would be most effective for signaling hormone-regulated noradrenergic-signaled information to the basal forebrain, for the purpose of behavioral controls or controls over the release of gonadotropin-releasing hormone (GnRH) release (Goubillon et al., 1999; Scott et al., 1999). In rats, as well, most noradrenergic terminals in the region of GnRH cell bodies in the basal forebrain originate in neurons located ipsilaterally in noradrenergic areas A1 or A2, the most caudal (Wright and Jennes, 1993). These projections could be important not only for stress signaling (Palkovits, Baffi, and Pacak, 1999) but also for courtship behaviors and for the release of GnRH (Clarke et al., 1999; Legan and Callahan, 1999).

What might be the functional importance of such neuronal connectivity from noradrenergic brain stem cell groups to the basal forebrain? On the one hand, it is clear that brain stem noradrenergic neurons can respond electrically to a variety of stimuli that should alert or arouse the animal. LC neurons, for example, are activated by a variety of internal (Valentino, Foote, and Aston-Jones, 1983; Valentino, Page, and Curtis, 1991; Lechner et al., 1997) and external stimuli (Jacobs et al., 1991) that would, in all cases, get the animal's attention. On the other hand, the projections and the functional importance of these neurons directly implicates them in the controls over the release of LH and reproductive behaviors. It has been known for many years that noradrenergic inputs to the basal forebrain regulate cyclic release of gonadotropin hormone

(Kordon et al., 1994; Herbison, 1997). While many of these actions on GnRH neurons are distinctly positive, ascending noradrenergic inputs may also participate in the fasting-induced suppression of the release of LH (Maeda et al., 1996).

At the behavioral level also the importance of noradrenergic inputs to the basal forebrain and hypothalamus are beyond question. Estrogen-dependent noradrenergic responses in the VMH have been linked indissolubly to the controls over lordosis behavior (Kow et al., 1991; Kow, Weesner, and Pfaff, 1992). That is, the administration of estrogen facilitates electrophysiological responses by hypothalamic neurons whose activation is required for female reproductive behavior. In parallel, noradrenergic inputs to the basal forebrain tend to inhibit sleep-active neurons and to excite waking-active neurons in a manner consistent with the hormone-dependent control, by that basal forebrain tissue, of a variety of courtship behaviors (Osaka and Matsumura, 1994).

## 2. Dopaminergic Systems

Dopaminergic systems ascending into the hypothalamus and POA have functional importance for neuroendocrine end points. Dopaminergic neurons in the arcuate nucleus of the hypothalamus and in the zona incerta, at least, are known to participate in the hypothalamic regulation of the release of pituitary hormones (McKenzie et al., 1984; James et al., 1987; Sanghera, Anselmo-Franci, and McCann, 1991). Likewise, hormone-dependent female courtship behaviors, as should be correlated with ovulation, depend on dopaminergic inputs. Motorically active female-typical appetitive behaviors are reduced by the loss of dopaminergic inputs to the basal forebrain (Caggiula et al., 1976), and the courtship behaviors (Fig. 14) of both female (Pfaus and Pfaff, 1992; Pfaus, Smith, and Coopersmith, 1999) and male rats (Hull et al., personal communication) are known to depend on dopamine.

With respect to their inputs, midbrain VTA dopaminergic neurons can display bursts of activity in response to brief auditory and visual stimuli, as should alert the animal (Horvitz, 1997). In addition, we note that dopaminergic VTA neurons can respond exquisitely to noxious stimuli. This is especially important because a major target point of midbrain dopaminergic neurons is the Nacc, recently implicated in pain control (Altier and Stewart, 1999). With respect to outputs, it is clear (for a review see Melis and Argiolas [1995]) that, while dopaminergic systems are not necessary for the performance of lordosis itself, the mesolimbic do-

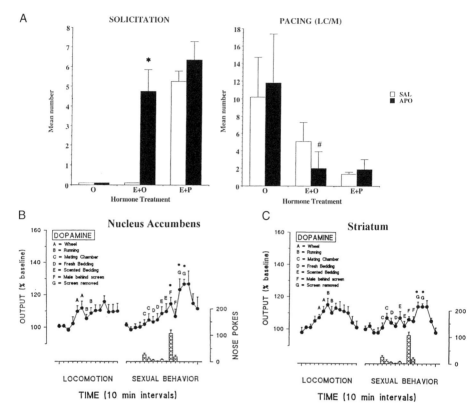

**Fig. 14.** Relations between the release of dopamine and courtship behaviors in female rats. (*A*) Apomorphine (APO), causing the release of dopamine, facilitates solicitation and reduces prolonged pacing (LC/M = level changes per mount) behaviors compared to the saline (SAL) control condition. Changes are especially evident under the estrogen-only (E+O) condition, showing that APO essentially puts the animal in a condition behaviorally similar to estrogen + progesterone (E+P). *, Significant apomorphine-induced increases in estrogen-activated solicitation behaviors in OVX rats; #, significant apomorphine-induced decreases in estrogen-inhibiting facing behaviors. (*B* and *C*) Relative to general locomotion, copulation (hatched bars) releases dopamine differentially in the nucleus accumbens (related to reward) compared to the striatum (related to general control of movements). *, Significant increases in dopamine levels in the nucleus accumbens (B) and striatum (C) following annotated behaviors. (From Pfaus, Smith, and Coopersmith, 1999.)

paminergic system plays a key role in the preparatory phases of reproductive behaviors, including sexual arousal and motivation.

It thus becomes most interesting that radioactive estrogens are bound in certain dopaminergic neurons (Sar, 1984; replicated by McCabe, Morrell, and Pfaff, unpublished observations) and, indeed, that sex steroids can promote neurite outgrowth in mesencephalic tyrosine hydroxylase immunoreactive neurons presumed to be dopaminergic neurons (Reisert et al., 1987). In cultured embryonic rat mesencephalic neurons, estrogen can upregulate tyrosine hydroxylase mRNA, particularly in cultures from female donors (Raab, Pilgrim, and Reisert, 1995). Moreover, in these cultures, numbers of tyrosine hydroxylase immunoreactive neurons develop sex-specific characteristics even in the absence of continuing sex differences in the hormonal environment. The rapidity of neurochemical changes in these neurons, together with the apparent absence of classical nuclear ERs, suggests that estrogens stimulate the differentiation of these dopaminergic neurons via a nongenomic signaling mechanism (Beyer, 1998). Indeed, a long history of experimental evidence gathered in the tradition of hormonal effects on instinctive behaviors (Becker and Beer, 1986; Mermelstein, Becker, and Surmeier, 1996; Xiao and Becker, 1998) shows that estrogens act in the striatum to modulate rapidly the release of dopamine and dopaminergic neural transmission in a manner not consistent with time-consuming genomic actions.

We surmise that sex hormones operate on dopaminergic systems important for the release of LH and reproductive behaviors by both genomic and nongenomic mechanisms. All of these data show that hormone-addressed dopaminergic neurons respond to a breadth of stimuli and have a range of outputs that qualify them for participation in general arousal mechanisms related to sex and pain.

As a side point, the evidence for estrogens addressing *histaminergic* systems can be stated quite simply. ER immunoreactivity is found in many histaminergic neurons (Fekete et al., 1999). Further, estradiol can affect certain histamine-evoked electrical responses in hypothalamic neurons (Jorgenson, Kow, and Pfaff, 1989).

### 3. Serotonergic Systems

The situation is somewhat different for serotonergic systems ascending from the raphe nuclei. Not only are there complexities with respect to distinctions between ER-α and ER-β, but also the tremendous variety of serotonergic receptor subtypes, defined pharmacologically and genetically, render functional predictions harder to sustain. On the one hand, the classical ER-α clearly is expressed within neurons of the rat DRN (Alves et al., 1998), but these neurons were not colocalized with the expression of tryptophan hydroxylase, indicating that ER-α and tryptophan hydroxylase are expressed within different cells in the same cell

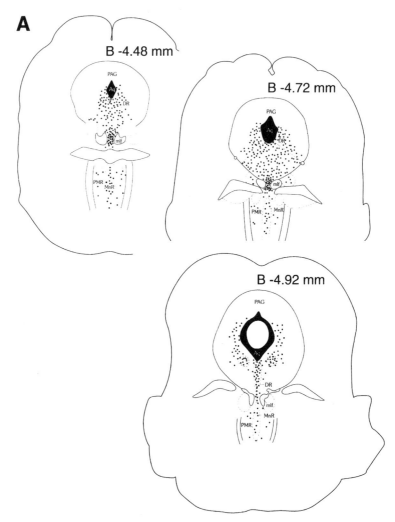

**Fig. 15.**   The distribution of cells exhibiting (*A*) progestin receptor (PR)-ir in a wild-type, estradiol benzoate–treated OVX mouse and (*B*) ER-α-ir in an OVX wild-type mouse. Sections are through the dorsal (DR) and median-paramedian raphe (MnR, PMR) nuclei and surrounding periaqueductal gray (PAG). Each dot represents one immunolabeled cell. Maps were drawn from actual 30-μm sections of representative animals. Note the greater density of PR-ir cells compared to ER-α-ir cells, as well as the abundance of cells colocalizing PR-ir and TPH-ir in the dorsal and median raphe nuclei. Aq., cerebral aqueduct; mlf, medial longitudinal fasiculus. (From Alves et al., 2001.)

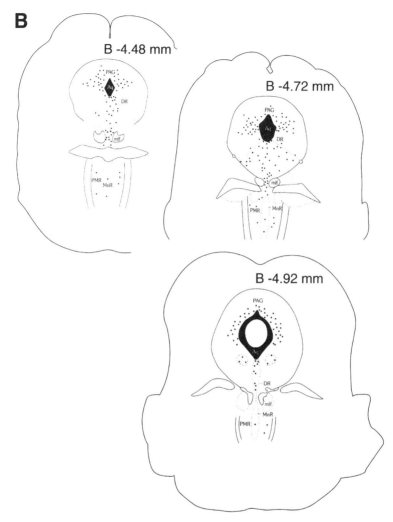

**Fig. 15.** (*Continued*)

group. ER-β immunoreactivity, also expressed in the DRN, may yield a different result with respect to serotonergic synthesizing mechanisms (Fig. 15) (Alves et al., 1999, 2001). Interestingly, in cats, neurons immunoreactive for the classical ER were seen in the vlPAG as well as in the rostral pole of the DRN (similar to rats; Pfaff and Keiner, 1973), and by a combination of immunocytochemistry with retrograde marker neuroanatomy, it was demonstrated that some of these ER-positive PAG neurons project to the medullary reticular formation (Vanderhorst et al., 1998).

In terms of endocrine importance, the release of serotonin plays an important role in the surge of LHRH crucial, in turn, for ovulation, working at least through the $5\text{-HT}_{2A}$ receptor (Fink et al., 1999). With respect to behavioral importance, it is well recognized that serotonergic systems are important for the control of mood and sleep. While decreased serotonergic activity is often associated with irritability, abnormally increased activity is often accompanied by undue sleepiness. Clearly, the effects of sex steroids on serotonergic neurons should be important for emotionally laden behaviors in animals and humans, in light of the overwhelming success of serotonin-active agents (e.g., selective serotonin receptake inhibitors [SSRIs]) as antidepressants (Fink et al., 1998). Thus, the effects of ovarian steroid on serotonin reuptake transporter mRNA (Pecins-Thompson, Brown, and Bethen, 1998) and serotonin-1A receptor mRNA (Pecins-Thompson and Bethea, 1998) in monkeys may be of special significance.

Nor should we forget autonomic adaptations to proceptive and receptive activities. ER immunoreactivity in autonomic neurons (Papka et al., 1997) and sensory neurons (cf. electrophysiological results: Kow and Pfaff [1973]) bespeaks the effects of hormones on autonomic controls both peripherally (Kauser and Rubanyi, 1998) and centrally (Rosas-Arellano, Solano-Flores, and Ciriello, 1999) that are important for reproductive behavior.

Finally, insofar as ascending monoaminergic systems subserving the effects of arousal also affect the ovulatory discharge of LH and female reproductive behaviors in a coordinate manner, they help achieve an adaptive synchrony between these endocrine and behavioral end points.

## D. Clinical Observations on Human Awareness and Arousal

Medical attention, by necessity, has concentrated on those conditions of human awareness and arousal notable by their very absence, namely coma and stupor (Plum and Posner, 1982). Coma, in which the patient is not aware of his or her environment and cannot be aroused by stimulation, and stupor, a condition requiring continuous stimulation for the patient to maintain even minimal levels of wakefulness or arousal, have been analyzed extensively to determine neural systems whose damage leads to the pathology (Plum, 1991). Clearly, bilateral damage to extensive regions of the upper mesencephalic tegmentum or bilateral destruction

**Table 2. Clinical Criteria for Patients in a Vegetative State**

Time durations
  One month, if persistent more than one year, almost always permanent
    No cognition: absence of consistent responses to linguistic, symbolic, or
      mimetic instruction
    No semantically meaningful sounds or goal-directed movements
    No sustained head-ocular pursuit activity

Functions usually or often preserved
  Brain stem and autonomically controlled visceral functions
    Homeothermia, osmolar homeostasis, breathing, circulation, gastrointestinal
      functions
    Pupillary and oculovestibular reflexes
    Brief, inconsistent shifting of head or eyes toward new sounds or sights
    Smiles, tears, or rage reactions
    Reflex postural responses to noxious stimuli

*Source:* Adapted from Plum et al. (1998, 56).

of the medial thalamus over a large anterior/posterior range will cause a vegetative or comatose state in humans. The vegetative state (Jennett and Plum, 1973) is characterized by preservation of hypothalamic and brain stem autonomic functions required to support minimal physical survival in the absence of awareness, self-awareness, or directed movements (Table 2). Occasionally, subconscious, modular physiological functions of forebrain tissue are evident (Plum et al., 1998). This state sometimes results from heroic medical procedures applied to patients with extensive brain damage. As an example of the neuronal loss that can be involved, Ingvar and Sourander (1970) reported a patient with severe damage to systems ascending from the mesencephalon and thalamus resulting from a serious brain stem hemorrhage. The patient at first spent several days in a comatose state and then "awakened" but remained permanently in a vegetative state.

Which neurochemical systems are involved in these disastrous conditions related to human consciousness? No single system dominates in the sense that its function is necessary and sufficient for consciousness. Instead, damage to ascending noradrenergic systems in the brain stem, ascending cholinergic systems originating in the lateral dorsal tegmentum of the midbrain and the upper pons, histaminergic systems originating in the mammillary bodies, serotonergic systems originating in the median raphe, and dopaminergic neurons projecting, in some cases, to the basal ganglia and in other cases to the limbic forebrain all can be involved. Damage to one or more of them can yield significant pathology

(Steriade, Contreras, and Amzica, 1997). The very multiplicity of ascending brain stem neurochemical systems contributing to different aspects of human awareness matches the results from animal studies that have used neuropharmacological tools to dissect five neurochemical systems contributing to alertness and arousal in experimental animals (references above).

From a clinical point of view, several components of human consciousness and their disturbances as a result of pathological conditions have been thoroughly charted (Table 3). Human awareness is not viewed as a unitary function (Posner, 1995). Systematically, dimensions of arousal and awareness have been explored. They include functions related to the alertness to sensory stimulation, the ability and inclination to initiate and sustain directed motor acts, and the ability to sustain a variety of emotional states, especially in anticipation of emotionally important events (Frohlich et al., 1999).

In summary (Plum, 1991), serious disturbances in human arousal and awareness result from bilateral and major destruction of the medial mesencephalic tegmentum or the medial thalamus. The data offer strong clinical support for the neurophysiological and neuroanatomical concepts of nonspecific arousal mechanisms arising in the brain stem and affecting wide areas of the forebrain. Functions lost in these pathological conditions include the ability to focus on inner or outer stimuli, initiate directed movements, and sustain a range of emotional states.

In certain cases, neurophysiological or neuroendocrine work carried out in animals has relevance for specific medical problems in humans separate from fundamental questions of awareness and arousal. Narcolepsy is a devastating condition recognized in human patients by sleepiness and a sudden loss of muscle tone that, for example, will cause a patient who is standing to fall down suddenly. Certain dogs exhibit narcoleptic-like symptoms and have been used for electrophysiological experiments. Siegel et al. (1991) identified neurons in the medial medulla of such dogs that fired at high rates only during cataplexy, the loss of antigravity muscle tone following emotional excitement and REM sleep. The latter is relevant, because it is thought that narcolepsy is a disease of REM sleep regulation (Siegel et al., 1991).

Other behavioral problems in humans are more closely related to the neuroendocrine themes emphasized in parts of this book. Cocaine addiction is thought to be particularly insidious in the manner in which this drug intercalates itself with dopamine receptors and dopaminergic

**Table 3. Major Components of Human Consciousness
and Their Clinical Disturbances**

Some clinical functions
   Activation of rostral mechanisms of arousal, psychological state functions, and
      cognitive integration
   Generation and regulation of psychological states of affect, mood, attention,
      and cathexis
   Generation and integration into consciousness relatively focal higher psycholog-
      ical functions of perception, memory, learned motor acts, and anticipation

Some clinical disturbances
   Disorders of arousal
      Coma
      Stupor
      Sleep disorders: hypersomnia, insomnia, narcolepsy

   Disorders of attention
      Distractibility, inattention, locked-on vigilance, obsessiveness

   Disorders of affect or emotion
      Anxiety or panic
      Agitation, irritability, lack of restraint, logorrhea, aggression
      Apathy, akinesia, mutism

   Disorders of psychic energy
      Indifference
      Fatigue syndrome and its congeners

   Disorders of global cognitive function
    Delirium and fugue states
      Multimodal dementia
      Vegetative state

   Impairments of focal conscious properties
      Agnosia
      Apraxia
      Aphasia
      Loss of anticipation
      Amnesia

*Source:* Adapted from Plum (1991, 360).

ascending systems related to reward, appetitive behaviors, and directed
movement in general. Quiñones-Jenab and colleagues (1996) examined
the effects of the administration of cocaine on stereotypic behavior and
locomotion in normally cycling female rats. Females in the estrous phase
of their cycle showed significantly higher cocaine-induced stereotypic
behavioral responses than those in other stages of the cycle. In terms of

locomotor activity, there was a strong interaction between the hormonal effects of estrus and the administration of cocaine, leading to very high levels of locomotor activity.

At the other end of the "excitability continuum" from vegetative states and coma—namely a state of hyperexcitability characterized by epileptic seizures—sex hormones may participate as important variables (Edwards, Burnham, and MacLusky, 1999; Edwards et al. 1999a, 1999b). Not only might epileptic seizures change around the time of puberty, but also women with catamineal epilepsy are more vulnerable to seizures when the ratio of circulating estrogens to progesterone is high. Under experimental conditions in which epileptic seizures are encouraged by a type of electrical stimulation called kindling, progesterone markedly elevates epileptic discharge thresholds. Progesterone slows the rate of seizure development and thus effectively has an anticonvulsive action. With other means of producing experimental seizures, as well, sex hormones alter the likelihood of catamineal epilepsy (Schwartz-Giblin, Korotzer, and Pfaff, 1989). Thus, aside from the histochemical and neuroanatomical evidence already cited, sex hormones apparently can change brain stem excitability in a fairly generalized fashion, in such a manner that mood and arousal could be affected.

## E. Mood

The emotional and autonomic effects of internal and external stimuli represent components in the structure of arousal and comprise one branch of the ascending reticular systems contributing to awareness. Sex hormones influence the neurochemical systems contributing to hypothalamic and limbic forebrain controls over autonomic and emotional states (see above). Some of the evidence simply shows hormonal effects on the norepinephrine, dopamine, and serotonin systems that project to hypothalamus and limbic forebrain. Other evidence directly demonstrates the effects of estrogens and progestins on mood.

Part of the work on how estrogens address noradrenaline synthesizing cells in the brain stem has been referred to above. Briefly, ERs clearly are identified in a subset of brain stem norepinephrine neurons, the groups called A1 and A2 in rats (Simonian and Herbison, 1997), and such findings have been extended to sheep (Scott et al., 1999). A direct molecular result of estrogen action in these cells is shown by their ability to synthesize progesterone receptor (Curran and Petersen, 1998; Bicknell,

1999). Fluctuations of ER and PR in these brain stem norepinephrine neurons during the rat estrous cycle suggest their physiological relevance (Haywood et al., 1999). Because these neurons project to the hypothalamus and basal forebrain, their likely role in arousal and mood is obvious.

Becker and others have shown dopaminergic systems also to be sensitive to estrogenic influences. The administration of estrogen increases the density of $D_1$ dopamine receptors in rat striatum (Hruska, 1986; Hruska and Nowak, 1988). Regarding the mesolimbic dopamine system, Reisert and colleagues at the University of Ulm have reported trophic-like sustaining effects of sex steroid on dopamine synthesizing neurons. Indeed, the rapid effects of progesterone on reproductive behaviors reported by Frye, Debold, and colleagues following implantation of progestins in the VTA could have their origins in mesolimbic dopamine neurons.

Long studied, the relation of serotonergic systems to mood has become a topic widely reported to the public following the overwhelming success of SSRIs as antidepressants (Barondez, 1993; Heisler et al., 1998; Ramboz et al., 1998). While the causal relations of individual serotonin receptor genes to emotional behaviors are far from being understood (see commentary in Julius [1998]), the clear involvement of serotonergic systems in the control of mood states makes it imperative to consider hormonal effects on these neurons and their receptors. Even some apparently unrelated effects of psychostimulants probably involve serotonergic systems (Gainetdinov et al., 1999). In part, we should consider serotonin-synthesizing cells in the raphe nuclei. In mice (Alves et al., 1999) and in monkeys (Bethea, 1993), a subpopulation of ER or PR immunoreactive cells, especially in the DRN, synthesize serotonin, as indicated by immunoreactivity for the primary synthesizing enzyme tryptophan hydroxylase. In addition, the many serotonergic receptor subtypes must be examined systematically. Importantly for female reproductive behavior, the administration of estrogen upregulates $5\text{-HT}_{1a}$ receptor expression in the VMH (Flugge et al., 1999). In addition, the results gathered by Fink and colleagues (Sumner and Fink, 1997, 1998; Fink et al., 1998, 1999; McQueen et al,. 1999) demonstrate an upregulation of $5\text{-HT}_{2a}$ receptors. Thus, in serotonergic systems, as in noradrenergic and dopaminergic, neurochemical evidence provides mechanisms for hormonal actions on mood.

Remember that changes in emotional state are highly correlated with changes in autonomic parameters, and causal effects work in both directions: influences of autonomic state on emotion and influences of emotion on autonomic state. In this connection, it is widely accepted that the administration of estradiol leads to vasodilatation (see review in Sarrel [1990]). Following administration of 17β-estradiol, systemic responses include increased cardiac output, increased heart rate, and decreased systemic vascular resistance. Mechanisms of hormonal action to explain this probably combine effects at peripheral sites with those in the CNS. While peripheral actions could include the kidney, most work in this domain focuses on direct estrogenic actions on blood vessels (see review in Mendelsohn and Karas [1999]). Rapid vasodilatation, clearly mediated by nongenomic mechanisms, occurs within a few minutes following estrogen exposure. Hormone-caused alterations in vascular reactivity (see review in White, Darkow, and Lang [1995]) could include alterations in reactivity to serotonin (Futo et al., 1992). Estrogen-caused vasodilatation could result directly through a cyclic guanosine 5′-monophosphate–dependent mechanism involving nitric oxide (White, Darkow, and Lang, 1995; Darkow, Lu, and White, 1997). Central mechanisms involving hormones are also possible. The baroreceptor reflex is regulated delicately by a circuit that includes the nucleus of the tractus solitarius and at least two cell groups in the ventrolateral medulla. In general, when sympathetic influences are predominant over parasympathetic influences, blood pressure is raised. However, the patterns of emotionally significant autonomic reactivity are much more complicated than a simple opposition between these two systems—individualized patterns of responsiveness of the two systems are much more important to consider. Intravenous injections of estrogens in rats can significantly enhance baroreceptor reflex sensitivity, for example, blocking the attenuation in baroreceptor reflex sensitivity observed after vagal stimulation (Saleh and Connell, 1998).

Do the effects of sex hormones play out in quantitative assays related to mood? Yes, absolutely. In the Porsolt forced swim test, one of the most heavily used assays for evaluating pharmacological treatments of "mood," an animal that struggles mightily and then "gives up" is an animal whose behavior is subject to improvement by antidepressants. Improvement comprises coordinated swimming as opposed to struggling or immobility. Dramatically, estrogen treatment of OVX female rats decreased time spent struggling and time of immobility (giving up), while correspondingly increasing the amount of time spent swimming (Rachman et al., 1998). On the second day of testing, the effect of estrogen was even more pronounced (Fig. 16) (Rachman et al., 1998). Likewise, in a series of experiments that also employed a chronic stress model of depression, the administration of estrogen attenuated immobility in depressed rats (Fiber and Etgen, 1998). Whether the mechanism of this large behavioral effect is due to direct actions of estrogens in the ascending noradrenergic or dopaminergic systems, as we have documented, remains a subject for future research.

But what about humans? Reduced estrogen levels in women have been shown to lead to depression, sleep disturbance, irritability, anxiety, and panic disorders (Campbell and Whitehead, 1977; Arpels, 1996; Sherwin, 1998). Clearly, in some women, administration of estradiol can reduce symptoms of depression (Rubinow, Schmidt, and Roca, 1998; Bloch et al., 2000) and increase a subjective sense of emotional energy and well-being (Justice and de Wit, 2000). By contrast, several days after the beginning of administration of progestin during sequential hormone replacement therapy (HRT), some women are beset by feelings of malaise (Yonkers, Bradshaw, and Halbreich, 2000). It is well understood that women are more frequently afflicted by depression than men (Bland, Orn, and Newman, 1988; Parry, 1989; Weissman, Livingstone, and Leaf, 1991; Ernst and Angst, 1992; Wittchen et al., 1992; Kessler et al., 1994). Among a subset of women, the prevalence of depression associated with large hormonal changes and reproductively significant events—postpartum depression and premenstrual dysphoric symptomology—might be more complex in their interpretation, although it seems likely that hormonal mechanisms are involved. Under circumstances in which HRT in elderly women has been shown to affect cognitive function, investigators (Hogervorst et al., 1998) have offered the interpretation that HRT has "global activating" effects, bearing on an enhancement of the speed of information processing. In a clinical study of postmenopausal women, estrogen treatment was associated with increased heart rate and increased respiratory amplitude and an altered parasympathetic responsiveness to stress, again suggesting that direct or indirect effects of hormones on autonomic functions could influence emotional status (Burleson et al., 1998).

Modern approaches to these clinical questions will expand on both molecular subjects and brain scanning. Regarding the former, the first and most obvious interpretation of steroid hormone actions involves transcriptional facilitation in the cell nucleus. However, the possibilities of rapid membrane action are also obvious, both for estrogens (Moss, Qin, and Wong, 1997) and for progestin metabolites that can act as anxiolytics through the benzodiazepine-sensitive portion of the $GABA_a$ re-

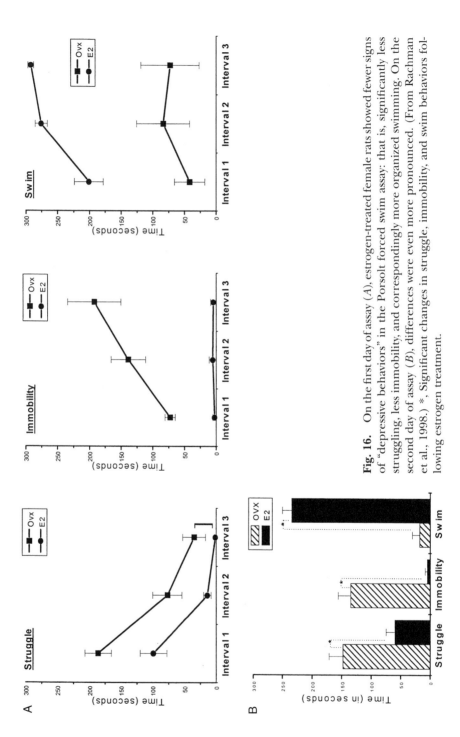

**Fig. 16.** On the first day of assay (*A*), estrogen-treated female rats showed fewer signs of "depressive behaviors" in the Porsolt forced swim assay: that is, significantly less struggling, less immobility, and correspondingly more organized swimming. On the second day of assay (*B*), differences were even more pronounced. (From Rachman et al., 1998.) *, Significant changes in struggle, immobility, and swim behaviors following estrogen treatment.

ceptor (Majewska et al., 1986; Freeman et al., 1993). Regarding the latter, positron emission tomography has already begun to be used to index regional cerebral blood flow during altered states of sexual desire, with an emphasis on the limbic forebrain (Stoleru et al., 1998).

## F. Summary

Neuronal systems that ascend in the brain stem and have widespread projections in the forebrain serve the most fundamental brain functions of arousal and awareness. Initially, they were studied and conceived as unitary and monolithic. Now they have been subdivided histochemically, neuroanatomically, and functionally. With respect to their neuroendocrine functions and sexual behavior, estradiol "activates" the female in a manner amplified by progesterone treatment. Elevated arousal and locomotor activity constitute only one of two motivational aspects of estrogenic action. The second aspect produces lordosis behavior itself. A novel concept thus emerged: that opposition between the two motivational mechanisms yields a biologically adaptive temporal sequence of behavioral performance (see also chapter I). In humans, the performance of these ascending systems, manifest in normal arousal and awareness, finds its clinical counterpart when one or more of the ascending systems is severely damaged, resulting in coma or stupor.

Still another hormonal effect is discussed in chapter III. Estrogenic modulation of opioid gene expression at a time when the female is receptive, will link sex, pain, and analgesia.

# III

## Descending Systems: The Importance of Opioid Peptides and Analgesia

### A. Opioid Peptides in Pain and Analgesia

Opioid peptides play crucial roles in central neural states that join sex systems and pain systems. Five classes of "opium-like" peptides and their corresponding receptors undisputedly help in the control of pain. Recently, by virtue of their hormone sensitivity, they have also been drawn into reproductive biology. Here, we talk about the reduction of pain (*analgesia*), the role of opioid peptides and their receptors in the molecular endocrinology of reproduction, and the necessity of pain reduction for reproductive behaviors. Notable is the participation of the vlPAG in analgesia, since that group of neurons represents an important intersection between lordosis behavior circuitry and analgesia pathways (see also section F).

### 1. Pain-Inhibitory Systems: Historical Overview of Central Morphine Analgesia and Stimulation-Produced Analgesia

Although opiates are among the oldest human analgesics, their central mode of action was discovered in 1964 in that profound analgesia was elicited by intracerebral morphine microinjection into the vlPAG (Tsou and Jang, 1964). Mapping studies revealed an analgesic rostrocaudal midline system responsive to morphine including the diencephalic, mesencephalic, and pontine central gray; the ventral thalamus; the DRN; and the nucleus raptic magnus (NRM) (Jacquet and Lajtha, 1973, 1974; Yaksh, Yeung, and Rudy, 1976). The nucleus reticularis gigantocelluaris (NRGC) and amygdala also supported morphine analgesia (Rodgers, 1977, 1978; Takagi et al., 1977). These analgesic responses were stereo-specific, were reversed by naloxone, displayed tolerance, and were cross-tolerant with systemically applied morphine.

Reynolds (1969) initially demonstrated that stimulation of the vlPAG produced profound analgesia, but the precise parameters were carefully defined by the research of John Liebeskind and colleagues. Thus, stimulation-produced analgesia (SPA) was elicited across mammalian species (Mayer et al., 1971; Liebeskind et al., 1973; Mayer and Liebeskind, 1974; Oliveras et al., 1974) and was soon demonstrated in terminally ill cancer patients for treatment of intractable pain (Adams, 1976; Hosobuchi et al., 1979). The time course of vlPAG SPA varies from seconds to hours in rat, cat, and monkey but can persist for days in humans. SPA was elicited from many of the sites that supported intracerebral morphine analgesia (Rhodes and Liebeskind, 1978; Oliveras et al., 1978) as well as the lateral hypothalamus, rubstantia nigra (SN), and LC (Balagura and Ralph, 1973; Rose, 1974; Segal and Sandberg, 1977), yet there was not a 1:1 relationship between sites supporting SPA and morphine analgesia (Mayer and Price, 1976; Lewis and Gebhart, 1977).

Central morphine analgesia and SPA shared some important similarities suggestive of common neural substrates (Yeung, Yaksh, and Rudy, 1977). Both display tolerance and partial cross-tolerance (Mayer and Hayes, 1975; Oliveras et al., 1978). Subanalgesic doses of morphine and sub-analgesic levels of stimulation summate to produce analgesia (Samanin and Valzelli, 1971). Both procedures selectively inhibited nociceptive-responsive (lamina V), but not nonnociceptive-responsive (lamina IV), cells of the dorsal horn of the spinal cord (Satoh and Takagi, 1971; Liebeskind et al., 1973; Oliveras et al., 1974; LeBars et al., 1975; Fields et al., 1977). Both vlPAG SPA and morphine analgesia, respectively, excite the NRM and inhibit nociceptive-sensitive neurons in laminae I, II, and V of the dorsal horn (Oliveras et al., 1974; Fields and Anderson, 1978; Behbehani and Pomeroy, 1978; Behbehani and Fields, 1979), yet morphine microinjected into the rostral ventromedial medulla (RVM) and vlPAG can also activate nociceptive dorsal horn neurons through an unknown, but possibly pain-facilitatory, process (Gebhart, 1986). Although morphine and electrical stimulation of the dorsolateral PAG elicit explosive motor behavior to light tactile stimuli (Mayer et al., 1971; Jacquet and Lajtha, 1974; Mayer and Liebeskind, 1974; Jacquet, Carol, and Russell, 1976), subsequent studies have indicated that the locus of analgesic action is in the vlPAG. Both responses are eliminated by lesions placed in either the vlPAG or the dorsolateral funiculus of the spinal cord (Samanin, Gulmulka, and Valzelli, 1970; Basbaum et al., 1977; Dostrovsky and Deakin, 1977; Rhodes, 1979), are reduced by either overall serotonin or dopamine depletion, yet are enhanced by overall noradrenergic depletion (Akil and Mayer, 1972; Vogt, 1974; Akil and Liebeskind, 1975; Hayes et al., 1977; Tenen, 1968; Yaksh and Rudy, 1978). These effects appear quite specific to the connection between the vlPAG and RVM, since stimulation of dorsal PAG sites produced responses not related to analgesia (Fardin, Oliveras, and Besson, 1984a, 1984b, 1984c).

Thus, these converging data from opiate microinjection and electrical stimulation studies allowed for the formulation of an intrinsic pain-inhibitory system, which soon was considered as a component of emotional and behavioral state change as proposed in the perceptual/ defensive/recuperative model (Bolles and Fanselow, 1980). Interrelationships of pain, analgesia, fear, and anxiety in motivational behavior were subsequently studied extensively (for reviews see Fanselow [1986]; Fanselow and Sigmundi [1987]; Fendt and Fanselow [2000]). Analgesic pain control can also be considered as a component of emotional state change related to aggression such as fight, flight, and freezing as well as recovery and healing (Bandler and Shipley, 1994; Bandler and Keay, 1996). Similarities exist between analgesic responses elicited by the vlPAG and other homeostatic responses such as arterial pressure, heart rate, regional vascular perfusion patterns, respiration, and neuroendocrine responses (e.g., Floyd et al., 1996; Keay et al., 1997; Henderson, Krey, and Bandler, 1998), leading to other conceptualizations of active and passive emotional states as well as vocalization behavior in which analgesia plays a role (for reviews see Davis, Zhang, and Bandler [1996]; Bandler, Price, and Keny [2000]).

As indicated previously for sexual behavior, the identification of an interconnected system confers particular advantages toward its further elaboration. Hence, one can study nociceptive inputs and then gauge behavioral, physiological, and functional changes in these inputs as a function of activating different aspects of the descending pain control system. In evaluating motivation in general, study of analgesia differs from other motivated states because of our considerable understanding of interconnections between stimulus input and response output, aided immeasurably by the conceptual and data-driven contributions of Alan Basbaum and Howard Fields.

## 2. A Proposed Endogenous Pain-Modulatory System

Basbaum and Fields (Fields and Basbaum, 1978; Basbaum and Fields, 1984) initially proposed a descending pain-inhibitory system consisting of three tiers: (1) the vlPAG, (2) the RVM, and (3) the dorsal horn of the spinal cord. In their model, opioids produce analgesia through inhibition in each structure. The nature of spinal inhibition is necessary and sufficient to produce analgesia through postsynaptic blockade of primary nociceptive afferents and excitatory interneurons in the substantia gelatinosa. However, supraspinal pain inhibition typically occurs through disinhibitory processes of the vlPAG and RVM, presumably through GABAergic interneurons. Thus, supraspinal opioid antinociception is mediated in part by neurons originating in the midbrain vlPAG, which synapse in the RVM (NRM, NRGC, NRGC pars $\alpha$) and which project to laminae I, II, and V of the dorsal horn, which give rise to ascending pain pathways. In addition to synapsing on these ascending nociceptive afferents, descending RVM neurons also synapse on cells containing $\mu$, $\kappa$, and $\delta$ opioid receptors (Fields, Heinricher, and Mason, 1991; Arvidsson et al., 1995a, 1995b; Mansour et al., 1995) and on enkephalinergic dorsal horn neurons (Cho and Basbaum, 1989; Glazer and Basbaum, 1984). Thus, electrophysiological responses in the vlPAG elicited by tail heating is inhibited by $GABA_A$ receptor antagonism, which is reversed by intrathecal $\mu$ antagonists, suggesting that spinal opioids produce $\mu$-mediated presynaptic inhibition of incoming spinal nociceptive afferents to the dorsal horn and thereby contribute to descending supraspinal nociceptive processing (Budai and Fields, 1998). Similarly, spinal $\alpha_2$-adrenoceptor antagonists reverse bicuculline-induced inhibition of heat-evoked responses in the vlPAG (Budai, Harasawa, and Fields, 1998). Detailed and extensive literature on the spinal analgesic actions of opiates can be found elsewhere (Yaksh and Rudy, 1978; Yaksh, 1984a, 1984b; Gebhart, 1986; Hammond, 1986).

Direct projections between the vlPAG and NRGC and between the vlPAG and NRM have been described (Abols and Basbaum, 1981; Beitz, Mullett, and Weiner, 1983; VanBockstaele, Pieribone, and Aston-Jones, 1989; Van Buckstaele, Aston-Jones, and Pierbone, 1991) that contain serotonin, enkephalins, neurotensin, and substance P (Beitz, 1982a, 1982b).

These vlPAG projections are anatomically and functionally differentiated from lateral, dorsal, and dorsolateral PAG projections (Bandler and Shipley, 1994; Beitz, 1995; Cameron et al., 1995). The vlPAG also projects to spinally projecting noradrenergic pontine and medullary cell groups (A5, A6, A7) (Clark and Proudfit, 1991a, 1991b, 1991c; Proudfit and Clark, 1991) with the densest vlPAG projections found in the A6 (LC) and A7 noradrenergic neurons (Ennis et al., 1991; Bajac and Proudfit, 1999). The RVM, in turn, sends physiologically relevant projections to the LC (Aston-Jones et al., 1986; Clark and Proudfit, 1991c).

The next section indicates that a further elaboration of endogenous pain control occurred with the identification of electrophysiologically defined cells within the RVM that, based on their response functions, not only predicted the presence of a noxious stimulus, but also a mechanism by which opiate agonists produced supraspinal control over noxious input.

### 3. On-Cells and Off-Cells in the RVM: A Neurophysiological Substrate

Two types of cells were identified in the RVM that provide a neural substrate for bidirectional modulation of nociceptive processing: off-cells and on-cells (Fields, Heinricher, and Mason, 1991). Off-cells cease activity immediately before the occurrence of nocifensive reflexes such as the tail-flick reflex, but are activated by systemic or central vlPAG administration of morphine (Fields et al., 1983; Cheng, Fields, and Heinricher, 1986), and are thought to serve as the pain-inhibiting output. On-cells increase their activity immediately before nocifensive tail-flick reflexes and are thought to exert a net facilitating effect on nociceptive processing. On-cell firing is inhibited by systemic and central administration of morphine (Barbaro, Heinricher, and Fields, 1986; Cheng, Fields, and Heinricher, 1986). Direct inhibition of on-cells and indirect opioid activation of off-cells are supported by depression of on-cells by iontophoretic application of morphine without alterations in off-cell activity (Heinricher, Morgon, and Fields, 1992). Further, glutamate-induced activation of medullary on-cells is inhibited by pretreatment with vlPAG, but not intrathecal morphine (Morgan, Heinricher, and Fields, 1992). Because medullary off-cells are not sensitive to direct opiate application, GABA has been proposed as a potential disinhibitory off-cell modulator in anatomical, physiological, and functional studies (Heinricher, Haws, and Fields, 1991; Heinricher et al. 1994; Skinner et al., 1997; Mitchell,

Lowe, and Fields, 1998). Off-cell activation by vlPAG opioids apparently involves both disinhibition and an *N*-methyl-D-aspartate (NMDA)-mediated positive feedback in the RVM: recruitment of an inhibitory input to on-cells and an excitatory connection to off-cells (Behbehani and Fields, 1979; Vanegas and Barbaro, 1984; Cheng, Fields, and Heinricher, 1986; Morgan, Heinricher, and Fields, 1992). This is consistent with the idea of parallel pain-facilitating on-cell and pain-inhibiting off-cell channels in both the PAG and RVM (see review in Heinricher and McGaraughty [1999]). Indeed, in vitro studies (Pan, Williams, and Osborne, 1990) of RVM slices reveal two major classes of neurons: one that is directly hyperpolarized by $\mu$ opioid agonists, and a second that receives a GABAergic inhibitory postsynaptic potential that is, in turn, inhibited by $\mu$ opioids to allow inhibition of spinal pain transmission (Pan and Fields, 1996).

A role for these RVM neurons has also been considered in supraspinal pain facilitation in a manner similar to that observed for facilitatory and inhibitory pain-modulatory systems (see reviews in Fields [1992, 2000]). Thus, on-cells are excited by noxious stimuli and thereby appear to have a pronociceptive function supported by morphine abstinence or acute physical dependence producing hyperalgesia during a period in which off-cells are inactive and on-cells are hyperactive (Bederson, Fields, and Barbaro, 1990). Furthermore, lidocaine administered into the RVM blocks this form of naloxone-induced hyperalgesia (Kaplan and Fields, 1991).

### 4. Opioid Peptides and Their Role in Supraspinal Analgesia

This section introduces the major families of opioid peptides that are derived from one of five gene precursor molecules—proopiomelanocortin (POMC), proenkephalin, prodynorphin, proorphanin-pronociceptin, and endomorphins (Table 4)—and their role in supraspinal analgesic processes.

#### a. Proopiomelanocortin

The POMC gene contains the 31–amino acid peptide, $\beta$-endorphin, as well as $\alpha$-and $\gamma$-endorphin, $ACTH_{18-39}$, $\alpha$-melanotropin, and corticotropin-like intermediate lobe protein (Mains, Eipper, and Ling, 1977; Roberts et al., 1979). Whereas the pituitary is the major site of POMC synthesis, the brain contains two distinct POMC-derived cell groups in the arcuate and surrounding periarcuate nuclei of the hypothalamus

**Table 4. Simplified Description of Interactions between Selected Opioid Peptides, Opioid Analogues, and Opiate Drugs, and Opioid Receptor Subtypes**

| *Opioid peptide:* | *Pharmacological action:* |
|---|---|
| β-endorphin | $\mu_1 \gg \mu_2 \sim \delta > \kappa_3 \ggg \kappa_1$ |
| Met-enkephalin | $\mu_1 > \mu_2 > \delta > \kappa_3 \ggg \kappa_1$ |
| Leu-enkephalin | $\mu_1 > \delta \gg \mu_2 \ggg \kappa_3 \ggg \kappa_1$ |
| Dynorphin A$_{1-17}$ | $\kappa_1 > \mu_1 > \mu_2 \gg \delta \gg \kappa_3$ |
| Dynorphin A$_{1-8}$ | $\kappa_1 = \delta > \mu_1 \ggg \mu_2 \gg \kappa_3$ |
| *Opioid analogue:* | *Pharmacological action:* |
| DAMGO | $\mu_1 > \mu_2 \gg \kappa_3 \ggggg \kappa_1 \gg \delta$ |
| U50488H | $\kappa_1 \ggggg \kappa_3 \sim \mu_1 \sim \mu_2 \sim \delta$ |
| NalBzOH | $\kappa_3 \sim \kappa_1 \sim \mu_1 \sim \mu_2$ |
| DPDPE | $\delta \ggg \mu_1 \gggg \kappa_1 \sim \kappa_3 > \mu_2$ |
| *Opiate drug:* | *Pharmacological action* |
| Morphine | $\mu_1 > \mu_2 \ggg \kappa_3 > \kappa_1 \gggg \delta$ |
| Ethylketocyclazocine | $\kappa_1 \sim \mu_1 \sim \mu_2 > \kappa_3 \gg \delta$ |
| Levallorphan | $\mu_1 > \kappa_1 \sim \mu_2 > \kappa_3 \gg \delta$ |
| Naloxone | $\mu_1 \gg \mu_2 > \kappa_1 > \kappa_3 \gggg \delta$ |

*Note:* Bold symbols = very high binding affinity for the peptide, analogue, and drug.

(Watson et al., 1978) and in the caudal region of the nucleus tractus solitarius (NTS) (Khachaturian et al., 1985). These extensive projections course into cell groups intimately involved in analgesic processes including the amygdala, PAG, NRM, NRGC, and LC (Khachaturian et al., 1985). Although the number of supraspinal structures that contains POMC appears to be restricted, the terminal zones of these opioid peptides are quite disparate, and it may not be accidental that POMC neuronal projection systems encompass many of the brain regions and pathways discussed in chapter II with respect to arousal.

Although β-endorphin analgesia (e.g., Loh et al., 1976; Tseng et al., 1979) was originally thought to be mediated through the μ receptor (Akil et al., 1984), subsequent studies clearly demonstrated that morphine and β-endorphin employ different anatomical and neurochemical pathways in exerting their supraspinal analgesic effects. Thus, it was shown that intrathecal naloxone blocked ventricular β-endorphin but not morphine analgesia (Tseng and Fujimoto, 1985), that immunoreactive spinal met-enkephalin is released following ventricular and intracerebral β-endorphin but not morphine (Tseng et al., 1985; Tseng and Wang, 1992), and that antibodies raised against met-enkephalin block β-endorphin but not morphine analgesia (Tseng and Suh, 1989). Spinal δ but not μ opioid receptors appear to modulate ventricular β-endorphin analgesia (Suh, Fujimoto, and Tseng, 1989). Ventricular morphine and β-endorphin only display additive analgesic effects following ventricular and intrathecal administration (Roerig, Fujimoto, and Tseng, 1988). Spinal cholecystokinin systems modulate ventricular β-endorphin but not morphine analgesia (Suh and Tseng, 1990, 1992; Tseng and Collins, 1992).

Within the vlPAG, barbiturate anesthesia respectively reduces and enhances morphine and β-endorphin analgesia (Smith, Robertson, and Monroe, 1992). vlPAG β-endorphin and morphine analgesia rely, respectively, on a spinal opioid component and on spinal adrenergic and serotonergic receptors (Suh, Fujimoto, and Tseng, 1989; Tseng and Tang, 1990; Tseng and Collins, 1991). Thus, intrathecal naltrexone maximally reduces vlPAG β-endorphin analgesia yet produces less marked effects on vlPAG morphine analgesia (Monroe, Smith, and Smith, 1997). By contrast, intrathecal serotonergic or $\alpha_2$-adrenergic antagonism maximally reduces vlPAG morphine analgesia but produces less marked effects on vlPAG β-endorphin analgesia. Although naltrexone and μ-selective antagonism blocked vlPAG morphine and β-endorphin analgesia, the slopes of the dose-inhibition curves were not parallel (Smith et al., 1992; Monroe et al., 1996), suggesting involvement of distinct receptor subpopulations.

## b. Proenkephalin

A second opioid gene product, proenkephalin, contains met-enkephalin and extended met-enkephalin peptides (Hughes et al., 1975; Kimura et al., 1980; Comb, Herbert, and Crea, 1982). Enkephalins are found primarily as interneurons in the telencephalon (e.g., cerebral cortex, olfactory tubercle, amygdala, hippocampus, bed nucleus of the stria terminalis and POA), diencephalon (e.g., hypothalamus, periventricular thalamus), mesencephalon (e.g., superior and inferior colliculi, PAG, interpeduncular nuclei), metencephalon-mylencephalon (e.g., LC, raphe nuclei, NRM, NRGC, NTS, lateral reticular nuclei), and spinal cord (dorsal and intermediate gray at all four spinal levels) (Hokfelt et al., 1977; Sar et al., 1978; Khachaturian et al., 1985). Longer extrinsic enkephalinergic pathways project from the amygdala to the PAG and adjacent dorsal raphe nuclei (Rizvi et al., 1991), as well as from the PAG to the NRM (Beitz, 1982a). Thus, in contrast to the larger opioid peptide, β-endorphin, the shorter enkephalin peptides appear to be far more ubiquitous in cell groups but, conversely, act mostly as intrinsic interneurons. Enkephalin analogues are discussed in subsequent sections to specify biochemical and analgesic actions of opioid receptor subtypes.

### c. Prodynorphin

The third major opioid gene system codes for preprodynorphin, which is cleaved into three leu-enkephalin-containing peptides: α-and β-neoendorphin, dynorphin A, and dynorphin B (Goldstein et al., 1981; Kangawa et al., 1981). Immunoreactive dynorphin-like perikarya are located in the telencephalon (cerebral cortex, striatum, amygdala, and hippocampus), diencephalon (supraoptic, paraventricular, and arcuate hypothalamic nuclei), mesencephalic (PAG), metencephalon-mylencephalon (parabrachial and spinal trigeminal nucleus, NTS, lateral reticular nucleus), and spinal cord (dorsal and ventral horns). Dynorphin A binding can occur with MOR-1, DOR-1, KOR-1, and KOR-3/orphanin-opioid receptor (ORL-1) receptor clones, which are described in detail in the next section (Zhang et al., 1998). Thus, the dynorphin gene products appear to have a pattern intermediate to that observed for β-endorphin and the enkephalins, and they possess both long fiber-projection systems and short intrinsic interneurons.

Although dynorphin peptides were initially proposed as potent opioid peptides, their analgesic effects were relatively poor in supraspinal (Friedman et al., 1981; Chavkin, James, and Goldstein, 1982) and spinal studies with nonopioid components of spinal dynorphin producing paraplegia and neurological deficits (Faden and Jacobs, 1983; Przewlocki, Shearman, and Herz, 1983; Herman and Goldstein, 1985; Caudle and Isaac, 1988; Long et al., 1988). Dynorphin $A_{1-8}$ analgesia can occur following coadministration of an endopeptidase 24.15 inhibitor (Chu and Orlowski, 1985) that was blocked by κ but not μ opioid antagonists (Kest, Orlowski, and Bodnar, 1992). It appears, however, that the levels of endogenous dynorphin peptides, particularly in the spinal cord, are increased in animal models of arthritis, inflammation, and chronic pain and may be related to alterations in nociceptive thresholds following such states (Millan et al., 1986, 1987, 1988). Indeed, κ receptor activation with dynorphin appears to act as an antianalgesic agent within the RVM (Pan and Fields, 1996) by inhibiting RVM off-cells, which, in turn, inhibit spinal pain transmission. By contrast, μ-selective agonists inhibit a separate subpopulation of RVM on-cells that facilitate pain transmission. Thus, whereas μ-mediated neurophysiological actions promote behavioral analgesia, dynorphin's κ-mediated neurophysiological actions promote antianalgesic responses.

## d. Proorphanin/Pronociceptin

Orphanin/nociceptin (OFQ/N) is a recently discovered heptadeca-peptide that is structurally similar to dynorphin A (Meunier et al., 1995; Reinscheid et al., 1995). Unlike classical opioid peptides, OFQ/N does not have a Tyr-Gly-Gly-Phe core, but rather a Phe-Gly-Gly-Phe motif. More-over, unlike traditional opioid peptides, it binds with very low affinity to classical opioid receptor subtypes. The preproorphanin-nociceptin gene contains additional pairs of basic amino acid residues that delineate two putatively biologically active peptides immediately downstream of OFQ/N, and it also contains a precursor peptide, nocistatin, that is also bio-logically active (Okuda-Ashitake et al., 1998). OFQ/N also has been localized in pain control nuclei including the amygdala, LC, PAG, raphe nuclei, and superficial layers of the dorsal horn (Henderson and McKnight, 1997; Letchworth et al., 2000). Within the spinal cord, the distribution of OFQ/N is highly similar to that observed for either enkephalin or dynorphin immunoreactivity (Riedl et al., 1996).

OFQ/N was so named because of its initial apparently paradoxical hy-peralgesic effects (Meunier et al., 1995; Reinscheid et al., 1995). How-ever, Mogil and coworkers found that supraspinal OFQ/N failed to alter nociceptive thresholds but blocked morphine and an opioid form of stress-induced analgesia (Grisel et al., 1996; Mogil et al., 1996a) as well as analgesia induced by $\mu$, $\delta$, and $\kappa$ opioid receptor agonists (Mogil et al., 1996b). By contrast, intrathecal OFQ/N potentiated morphine anal-gesia (Tian et al., 1997). However, the hyperalgesic and potential anti-opioid actions of OFQ/N appear to be strictly dependent on stimulus and temporal parameters such that spinal OFQ/N can produce naloxone-reversible analgesia (Rossi, Leventhal, and Pasternak, 1996; King et al., 1997), and that OFQ/N, $OFQ/N_{1-7}$ and $OFQ/N_{1-11}$ in the amygdala each produced analgesia but not hyperalgesia on the tail-flick, but not the jump, test (Shane et al., 2000). This pattern is dramatically different from analgesia elicited by either morphine, $\beta$-endorphin, or U50488H in the amygdala which is most pronounced on the jump test and less pro-nounced on the tail-flick test (Rodgers, 1977, 1978; Pavlovic, Cooper, and Bodnar, 1996a; Pavlovic and Bodnar, 1998).

The precursor gene for OFQ/N has other active compartments, par-ticularly $OFQ_{160-187}$, which produces analgesia following ventricular ad-ministration (Rossi et al., 1998), and intracerebral administration in the amygdala, vlPAG, LC, and RVM (Shane et al., 1999). The amygdala and vlPAG displayed naltrexone-reversible $OFQ/N_{160-187}$-induced analgesia but hyperalgesia. By contrast, the LC and RVM displayed naltrexone-insensitive $OFQ/N_{160-187}$-induced analgesia and $OFQ/N_{160-187}$-induced hyperalgesia. Such double dissociations ensure specificity of effects within sites and indicate that the multivariate pronociceptive and antinociceptive actions of OFQ/N may indeed be site specific. The neurophysiological correlates of OFQ/N analgesia and hyperalgesia were identified in the RVM such that OFQ/N produces hyperalgesia by antagonizing μ-mediated opioid analgesia and produces analgesia during opioid abstinence. OFQ/N apparently produces both effects by inhibiting, respectively, on-cells involved in pain facilitation and off-cells involved in pain inhibition (Pan, Hirakawa, and Fields, 2000).

### e. Endomorphins

Endomorphins are the most recent opioid peptides to be isolated from the brain (Zadina et al., 1997). Like OFQ/N, these tetrapeptides differ in their core from classical opioid peptides: endomorphin-1 (Tyr-Pro-Trp-Phe-$NH_2$) and endomorphin-2 (Tyr-Pro-Phe-Phe-$NH_2$). Immunohistochemical evidence revealed that endomorphins are present in the dorsal horn of the spinal cord and a number of medullary structures as well as in the PAG, LC, and amygdala (Martin-Schild et al., 1997, 1999). Whereas endomorphin-1 is more prevalent in the brain, endomorphin-2 is more prevalent in the spinal cord. Thus, endomorphins share a number of spinal areas relevant to analgesic responses with the classical opioid peptides, particularly dynorphins and enkephalins.

Endomorphin-1 and endomorphin-2 produce analgesia sensitive to general and μ opioid antagonism (Zadina et al., 1997). Both endomorphin-1 and endomorphin-2 analgesia occur following spinal and supraspinal administration that is blocked by μ and $μ_1$ opioid antagonists, yet CXBK mice deficient in μ opioid receptors fail to display endomorphin-1 or endomorphin-2 analgesia (Stone et al., 1997; Goldberg et al., 1998; Tseng et al., 1998; Sakurada et al., 1999). Whereas endomorphin-1-induced analgesia is selectively blocked by μ but not δ or κ opioid antagonists, endomorphin-2-induced analgesia is blocked by both μ and κ opioid antagonists as well as antisera directed against dynorphin $A_{1-17}$ (Tseng et al., 1998). Endomorphin-2-induced analgesia is also modulated by the endopeptidase, dipeptidyl peptidase IV, and its resistant analogue, D-Pro²-endomorphin-2, produces more potent and longer-lasting analgesia following ventricular administration (Shane, Wilk, and Bodnar, 1999b).

## 5. *Opioid Receptor Subtypes and Supraspinal Analgesia*

Following the discovery of the opiate receptor in 1973 (Pert and Snyder, 1973; Simon, Hiller, and Edelman, 1973; Terenius, 1973), biochemical and physiological evidence was provided for the existence of multiple receptor subtypes (Martin et al., 1976; Lord et al., 1977) (Table 4). Although three distinct opioid receptors were initially proposed based on the lack of cross-tolerance among prototypical agonists—μ (morphine), κ (ketocyclazocine), and σ (SKF10047) (Martin et al., 1976)—the latter was subsequently characterized as nonopioid (Vaupel, 1983). The δ receptor was described as an enkephalin-preferring receptor because enkephalins were more potent than morphine in the mouse vas deferens assay (Lord et al., 1977).

Localization of endogenous opioid peptides and receptors revealed that there is relatively poor anatomical correspondence between them (Akil et al., 1984). Further, there appears to be crossreactivity between opioid peptides and opioid receptors in binding assays (Table 4). Thus, β-endorphin selectively binds both μ and δ, but not κ, receptors. Enkephalins and dynorphins bind preferentially to δ and κ receptors, respectively, in vitro. Moreover, all proenkephalin and prodynorphin peptides can bind to μ, δ, and κ opioid receptors depending on the peptide product and species (Corbett et al., 1982; Quirion and Weiss, 1983). Unlike classical opioid peptides, the newer opioid peptides demonstrate selective biochemical correspondence with their endogenous opioid receptors. Hence, whereas OFQ/N shows little or no affinity for μ, δ, and κ opioid receptors, it displays high affinity for the orphanin receptor (Meunier et al., 1995; Reinscheid et al., 1995). Endomorphin-1 and endomorphin-2 display high affinity and selectivity for the μ receptor, and each has been proposed as the actual endogenous ligands for this receptor subtype (Zadina et al., 1997). The following sections summarize the major μ, δ, and κ opioid receptor subtypes and their pharmacological and biochemical characterization.

### a. μ *Opioid Receptors*

The μ receptors are widely distributed throughout the brain and especially in areas related to pain control including the amygdala, PAG, dorsal raphe, LC, NRM, NRGC, and dorsal horn of the spinal cord (Mansour et al., 1988). The μ receptor has been characterized pharmacologically using selective agonists such as D-Ala$^2$, Met-Phe$^4$, Gly-ol$^5$-encephalin (DAMGO) and antagonists such as β-funaltrexamine (β-FNA) and Cys$^2$-Tyr/3-Orn$^5$-Pen$^7$ (CTOP). It has been classified further into $\mu_1$ and $\mu_2$ receptor subtypes based on pharmacological assays in which naloxonazone and naloxonazine selectively antagonize $\mu_1$ receptor actions in vitro and in vivo (for a review see Pasternak and Wood [1986]). The $\mu_1$ receptor binds opiates and most enkephalins with similar high affinity, while the $\mu_2$ receptor binds morphine more potently than enkephalins. Autoradiographic studies revealed similar, but not identical, distributions of $\mu_1$ and $\mu_2$ receptors (Goodman and Pasternak, 1985; Moskowitz and Goodman, 1985). Behavioral studies have distinguished the actions of $\mu_1$ and $\mu_2$ receptors in spinal and supraspinal analgesic processes (Bodnar et al., 1988; Paul and Pasternak, 1988; Paul et al., 1989) and in ingestive behavior (Bodnar, 1996). Jensen and Yaksh (1986a, 1986b, 1986c) compared the abilities of selective $\mu$ and $\delta$ opioid agonists to produce analgesia in the vlPAG and RVM and found that $\mu$ receptor agonists were far more effective across a range of pain tests (Table 5). These effects were confirmed by others as well in the NRM, NRGC, and NRGC pars $\alpha$ (Takagi et al., 1977; Akaike et al., 1978; Azami, Llewelyn, and Roberts, 1982; Schmauss and Yaksh, 1984). Several investigators suggested that based on agonist studies, the $\mu$ receptor was the integral opioid receptor for analgesia in the vlPAG and RVM (Fang, Fields, and Lee, 1986; Leander, Gesellchen, and Mendelsohn, 1986; Smith et al., 1988). Using selective antagonists, $\mu_1$ sites have been implicated in several opiate actions, including supraspinal intracerebroventricular analgesia (Table 6), but not others, such as respiratory depression, inhibition of gastrointestinal transit, and many signs of physical dependence (Paternak, Childers, and Sny-

**Table 5. Summary of Opioid Receptor Subtype Agonists and Intracerebral Analgesic Activity**

| Site | Morphine | U50488H | DPDPE | Deltorphin |
|------|----------|---------|-------|------------|
| ICV | +++ | ++ | ++ | ++ |
| vlPAG | +++ | None | None | ++ |
| RVM | +++ | None | None | ++ |
| LC | +++ | None | None | ? |
| Amygdala | +++ | ++ | ++ | ++ |

ICV, intracerebroventricular; vlPAG, ventrolateral periaqueductal gray; RVM, rostral ventromedial medulla; LC, locus coeruleus; DPDPE, D-Pen$^2$, D-Pen$^5$-enkephalin. +++, potent analgesic response; ++, moderate analgesic response; none, no analgesic response; ?, not tested.

**Table 6. Selective Opioid Receptor Subtype Anatagonists and Antisense Oligodeoxynucleotide Effects on Opioid-Mediated Analgesic Responses**

| Antagonist | Subtype | Mor | M6G | Heroin | U50488 | DPDPE | Delt II |
|---|---|---|---|---|---|---|---|
| Ntx | General | Yes | Yes | Yes | Yes | Yes | Yes |
| β-FNA | μ | Yes | Yes | Yes | No | No | No |
| Naz | $\mu_1$ | Yes | Yes | Yes | No | No | No |
| NTI | δ | No | No | No | No | Yes | Yes |
| NBNI | $\kappa_1$ | No | No | No | Yes | No | No |

| AS ODN | Exon | Mor | M6G | Heroin | U50488 | DPDPE | Delt II |
|---|---|---|---|---|---|---|---|
| MOR-1 | 1 | Yes | No | Yes | No | No | No |
| MOR-1 | 2 | No | Yes | Yes | — | — | — |
| MOR-1 | 3 | No | Yes | Yes | — | — | — |
| MOR-1 | 4 | Yes | No | No | — | — | — |
| DOR-1 | 1 | No | No | No | No | Yes | Yes |
| KOR-1 | 1 | No | No | No | Yes | No | No |

Mor, morphine; M6G, morphine-6β-glucuronide; U50488, $\kappa_1$ agonist; DPDPE, $\delta_1$ agonist; Delt II, deltorphin ($\delta_2$ agonist); Ntx, Naltrexone; β-FNA, β-funaltrexamine (μ antagonist); Naz, naloxonazine ($\mu_1$ antagonist); NTI, naltrindole (δ antagonist); NBNI, nor-binaltorphamine ($\kappa_1$ antagonist); AS ODN, antisense oligodeoxynucleotide; MOR-1, μ opioid receptor clone; DOR-1, δ opioid receptor clone; KOR-1, κ opioid receptor clone.

der, 1980; Spiegel, Kourides, and Pasternak, 1982; Ling and Pasternak, 1983; Ling et al., 1984, 1986).

In assessing the ability of opioid receptor subtype agonists to elicit analgesia from the vlPAG, RVM, and LC, morphine and the enkephalin derivative D-Ser[2], Leu[5]-enkephalin (DSLET) produced marked analgesia, which was blocked by naloxonazine (Bodnar et al., 1988) (Table 5). DSLET's elicitation of $\mu_1$-mediated analgesia correlates with its ability to bind $\mu_1$ receptors (Itzhak and Pasternak, 1987). The putative κ opiate, ethylketocyclazocine, produces analgesia following ventricular administration that is blocked by $\mu_1$ antagonism and binds $\mu_1$ receptors with high affinity (Pasternak, 1980; Ling and Pasternak, 1983). Although ethylketocyclazocine failed to elicit analgesia from either the vlPAG or the LC alone, it significantly reduced the analgesic actions of coadministered morphine or DSLET into the same structure (Smith et al., 1988; Bodnar, Paul, and Pasternak,, 1991), suggesting potential action as a partial μ agonist, producing minimal subthreshold analgesic effects at each site alone, and preventing morphine or DSLET from producing maximal effects at the relevant receptors. To test this hypothesis, simultaneous administration of ethylketocyclazocine into the vlPAG and LC produced naloxonazine-reversible analgesia, indicating the importance of regional interactions to produce partial agonist effects. Thus, it appears that $\mu$ receptors are extremely important in mediating supraspinal opioid analgesia.

### b. δ Opioid Receptors

Autoradiographic studies indicate that δ receptor binding is densest in the olfactory tubercule, neocortex, neostriatum, Nacc, and amygdala (Mansour et al., 1988). General δ receptor agonists include DSLET and DADL, while general δ receptor antagonists include ICI174864 and naltrindole (NTI). The δ receptor has also been classified further into $\delta_1$ and $\delta_2$ receptor subtypes, with the former pharmacologically characterized by the agonist actions of D-Pen$^2$, D-Pen$^5$-enkephalin (DPDPE) and longer-term antagonist actions of D-Ala$^2$, Leu$^5$, Cys$^6$-enkephalin. The $\delta_2$ receptor has been pharmacologically characterized by the agonist actions of D-Ala$^2$, Glu$^4$-deltorphin (DELT) and the antagonist actions of NTII. Behavioral studies also have distinguished the actions of $\delta_1$ (DPDPE) (Porreca et al., 1984; Heyman et al., 1988) and $\delta_2$ (deltorphin) receptors in spinal and supraspinal analgesia (Jiang et al., 1991; Mattia et al., 1992). However, $\delta_1$ agonists fail to elicit analgesia from the vlPAG, LC, or RVM (Bodnar et al., 1988) (Table 5).

### c. κ Opioid Receptors

The κ receptor binding is densest in the striatum, nucleus accumbens, amygdala, hypothalamus, neurohypophysis, median eminence, and NTS (Mansour et al., 1988). Selective agonists and antagonists have also distinguished multiple κ receptor subtypes. The $\kappa_1$ receptor subtype has been characterized using the agonist U50488H (VanVoigtlander, Lahtir, and Ludens, 1983), and the antagonist nor-binaltorphamine (NBNI). Ventricular administration of U50488H produces analgesia, but direct administration into the vlPAG and LC does not (Czlonowski, Millan, and Herz, 1987; Bodnar, Paul and Pasternak, 1991). The $\kappa_2$ receptor subtype has been described in biochemical assays as being U50488H insensitive but has not been demonstrated in vivo (Zukin et al., 1988). A $\kappa_3$ receptor also has been identified as a U50488H-insensitive and NBNI-insensitive site that selectively binds the agonist naloxone benzoylhydrazone (NalBzOH). NalBzOH induces analgesic responses that are blocked by general opioid antagonists, but not NBNI (Gistrak et al., 1989; Paul et al., 1990).

### 6. Opioid Receptor Genes and Antisense Oligodeoxynucleotide Probes in Analgesic Processes

Although opioid receptors were the first class of receptors to bind bioactive peptides, they were among the last class of receptors to be cloned. In 1992, expression cloning yielded the amino acid sequence of the $\delta$ opioid receptor gene (DOR-1), which provided the first molecular evidence for the existence of opioid receptors (Evans et al., 1992; Kieffer et al., 1992). Subsequently, several groups successfully cloned the cDNAs encoding the $\mu$ (MOR-1), $\kappa$ (KOR-1), and $\kappa_3$ (KOR-3) receptors (Uhl, Childers, and Pasternak, 1994). Initially, the KOR-3 gene was thought to be the $\kappa_3$ receptor, but it was later shown to exhibit a high degree of sequence homology to the ORL-1 gene that was subsequently referred to as the KOR-3/ORL-1 gene. Each of the opioid receptor genes encodes a G-protein-coupled receptor ($G_I/G_o$), and shares homology of 65–70 percent of their amino acid sequence, primarily in their transmembrane-spanning regions and intracellular loops (Reisine and Bell, 1993). The DOR-1, KOR-1, and KOR-3/ORL-1 genes contain three coding exons, whereas the MOR-1 gene encodes four exons.

### a. DOR-1 Gene

The DOR-1 gene displayed a much higher affinity for enkephalin peptides and $\delta$ receptor agonists and antagonists than for dynorphin, $\mu$ and $\kappa$ agonists, and $\mu$ antagonists (Evans et al., 1992; Kieffer et al., 1992). In situ hybridization and immunohistochemical techniques initially identified the DOR-1 receptor gene on primary afferent nerve terminals in the superficial layers of the spinal cord (Dado et al., 1993). The DOR-1 gene and its receptors are found presynaptically to regulate the release of transmitters from small-diameter primary afferent neurons. In the brain, DOR-1 immunohistochemistry reveals high concentrations that are highly similar to those observed for the $\delta$ receptor in autoradiographic studies (Arvidsson et al., 1995a, 1995b; Mansour et al., 1995). A gene knockdown technique using antisense oligodeoxynucleotides (AS ODNs) (Wahlestedt, 1994) can correlate the activity of these genes with in vivo opioid receptor pharmacology (Pasternak and Standifer, 1995) (Table 6). Thus, AS ODN probes directed against the DOR-1 gene selectively blocked spinal analgesia following DPDPE ($\delta_1$) and deltorphin ($\delta_2$) in rats (Standifer et al., 1994), correlating well with the downregulation of mRNA and $\delta$ receptor protein (Standifer et al., 1995; Lai et al.,

1996). These initial studies examining DOR-1 have been confirmed and extended by others (Bilsky et al., 1994; Lai et al., 1994a, 1994b; Tseng, Collins, and Kampine, 1994), and AS ODN probes directed against the DOR-1 gene block increased morphine sensitivity following chronic naltrexone treatment (Kest et al., 1997).

### b. MOR-1 Gene

A gene encoding the μ opioid receptor (MOR-1) was isolated from rat and human that displays high affinity for μ-selective agonists (DAMGO) and antagonists (β-FNA, naloxonazine), and low affinity for δ (DPDPE) and κ (U50488H) agonists (see review in Uhl, Childers, and Pasternak, [1994]). In situ hybridization and immunohistochemical localization have identified the MOR-1 receptor gene in such pain-control nuclei as the amygdala, LC, NTS, NRM, and raphe nuclei (Schulz et al., 1998), which correlates well with autoradiographic studies using μ ligands (Arvidsson et al., 1995a, 1995b; Mansour et al., 1995).

Using homologous recombination techniques, the μ opioid receptor gene can be disrupted, resulting in an animal with no overt behavioral abnormalities. Yet, this mouse model displays an absence of morphine analgesia with largely intact responses to δ and κ opioid ligands (Matthes et al., 1996, 1998). This functional and selective absence of μ-mediated analgesia was accompanied by a selective loss of μ but not δ or κ opioid receptor autoradiography (Kitchen et al., 1997). Most important, mice lacking exon 1 or the MOR-1 clone fail to display morphine analgesia, but possess intact analgesic responses to heroin and the morphine metabolite, morphine-6β-glucuronide (M6G), indicating the presence of a unique site of action for these latter μ-mediated responses (Schuller et al., 1999). AS ODN probes directed against the MOR-1 gene selectively block μ-mediated analgesia (Rossi et al., 1994, 1995; Chen et al., 1995; Rossi, Standifer, and Pasternak, 1995). AS ODN approaches have been used successfully against individual exons of the mRNA encoding opioid receptors (Standifer et al., 1994). Thus, morphine analgesia is blocked by AS ODN probes targeting either exon 1 or exon 4 of the MOR-1 gene, but not those directed against exons 2 or 3 (Rossi et al., 1995, 1997b; Rossi, Standifer, and Pasternak, 1995). M6G is a very potent analgesic, 100-fold greater than morphine (Pasternak et al., 1987; Abbott and Palmour, 1988; Sullivan et al., 1989). However, M6G labels the traditional μ receptor slightly less potently than morphine (Paul et al., 1989), which brings into question how M6G possesses greater analgesic potency. The knockdown technique revealed a possible answer in that AS ODN probes directed against exons 2 or 3 blocked M6G analgesia, whereas probes directed against exons 1 or 4 were ineffective (Table 6), suggesting the existence of distinct morphine and M6G receptors in mice and rats that may represent splice variants of the MOR-1 gene. Similarly, AS ODN probes directed against the $G_{i\alpha2}$ subunit lowered morphine, but not M6G analgesia, whereas the loss of $G_{i\alpha1}$ blocked M6G analgesia, but not morphine analgesia (Rossi et al., 1995; Rossi, Standifer, and Pasternak, 1995). Finally, heroin analgesia was blocked by AS ODN probes directed against exons 1 and 2 of the MOR-1 gene (Rossi et al., 1996) (Table 6).

### c. KOR-1 Gene

The κ opioid receptor (KOR-1) was cloned from rodents and shows high affinity for dynorphin, U50488H, and NBNI, but low affinity for enkephalins (see review in Uhl, Childers, adn Pasternak [1994]). In situ hybridization has identified KOR-1 receptor mRNA in pain-control nuclei including amygdala, PAG, and LC, which correlates moderately well with previous autoradiographic studies (Mansour et al., 1995). Mice made deficient by homologous recombination of the KOR-1 gene display absences in $\kappa_1$ receptor binding but display no losses in either μ or δ opioid receptor binding (Slowe et al., 1999). AS ODN probes directed against the KOR-1 gene selectively block $\kappa_1$-mediated analgesia (Adams et al., 1994; Chien et al., 1994; Pasternak et al., 1999).

### d. KOR-3/ORL-1 Gene

A previously unrecognized $\kappa_3$ and orphanin-opioid receptor (KOR-3/ORL-1) gene was the fourth member of the opioid receptor family to be cloned (e.g., Bunzow et al., 1994; Mollereau et al., 1994; Pan, Cheng, and Pasternak, 1994; Pan et al., 1995). While OFQ/N, the endogenous ligand for the receptor, binds with high affinity to the KOR-3/ORL-1 gene (Meunier et al., 1995; Reinscheid et al., 1995), DAMGO, U50488H, and DPDPE were all inactive in assays involving the KOR-3/ORL-1 gene, indicating its low affinity for traditional opioids and opiates (Pan et al., 1995). In situ hybridization and immunohistochemical localization revealed KOR-3/ORL-1 receptor mRNA in the amygdala, PAG, raphe nuclei, and LC (Anton et al., 1996). AS ODN probes directed against the KOR-3 gene selectively block OFQ/N-mediated and $\kappa_3$-mediated analgesia (Pan, Cheng, and Pasternak, 1994). Hyperalgesia elicited by OFQ/$N_{1-17}$ and OFQ/$N_{1-11}$ is blocked by an AS ODN probe directed against exon 1 but not exons 2 or 3 of the KOR-3/ORL-1 gene, whereas analgesia elicited by these peptide fragments is blocked by AS ODN probes directed against exons 2 or 3, but not exon 1 of the same gene (Rossi et al., 1997a).

Thus, the five identified gene-related opioid peptide families appear to interact with multiple opioid receptors that have been identified using classical biochemical and pharmacological techniques. The successful cloning of the opioid receptor genes has provided converging evidence for the multiplicity of these receptors. The functional significance of the opioid peptides and the multiple receptors at which they act and interact encompass a wide array of behavioral, physiological, and homeostatic states.

## 7. Strain Differences in Opioid and Opioid-Mediated Analgesia in Mice

Many of the aforementioned studies typically used the Sprague-Dawley rat or outbred mouse strains. However, recent studies using murine genetics have uncovered differences in analgesic responsivity as a function of strain. This section briefly reviews these findings. The major types of mice typically tested for morphine analgesia include the DBA/2 and C57BL/6 inbred strains; the recombinantly inbred CXBK strain, which is relatively insensitive to opiate analgesia; and mouse lines selectively bred for analgesia induced by the potent $\mu$ opioid agonist, levorphanol (HAR and LAR). The DBA/2 mice display greater morphine analgesia on the hot-plate test than the C57BL/6 strain, while the CXBK strain fails to display morphine analgesia. However, the patterning, intensity, and potency of analgesia in these groups changed as a function of nociceptive tests including the tail-flick test, hot-plate test, acetic acid writhing test, and formalin test. LAR strains displayed significantly greater morphine analgesia on the hot-plate and writhing tests than HAR strains (Mogil et al., 1996c). Genotypic differences can also be observed for inflammatory forms of nociception in which there are both sensitive (e.g., C57BL/6J) and insensitive (A/J) strains of mice (Mogil, Lichtensteiger, and Wilson, 1998). In addition to the inability of CXBK mice to display normal morphine analgesia, CXBK mice display decreases in $\delta_2$-mediated analgesic responses, opioid binding, and mRNA levels of the DOR-1 clone relative to BALB/c mice (Kest et al., 1998a). Whereas HAR-bred mice display more marked increases in potency to levorphanol and DAMGO analgesia than LAR-bred mice, the two groups failed to display changes in analgesia elicited by $\delta_1$ and $\delta_2$ agonists (Kest et al., 1999). Greater morphine dependence is observed in mice bred for low (LAR) relative to high (HAR) analgesic responses to levorphanol (Kest et al.,

1998b). In testing 11 different strains of mice on 12 different nociceptive tests, strain-dependent and test-dependent differences in nociceptive responses ranged from 1- to 54-fold in sensitivity with a relatively high level of heritability across generations of these mice (Mogil et al., 1999a, 1999b). Thus, this variable is receiving increased attention in evaluating the underlying circuitry and pharmacology of opioid analgesia.

As indicated throughout the book, however, the majority of work especially examining intracerebral sites of action of analgesic processes has been performed in the rat. Therefore, in the following section, we return to that species to fully characterize μ receptor–mediated interactions between sites in supraspinal opioid analgesia.

## 8. Interactions between Supraspinal Opioid Analgesia Sites: Antagonist Studies

Neuronal cell groups in the brain stem related to the control of pain do not necessarily act by themselves. Instead, there are important functional interactions among these neuronal groups that are essential for us to understand, if we are to explain the neurophysiological and neurochemical basis of analgesia. A first approach to establish regional interactions between supraspinal sites in opioid analgesia is to determine whether opioid analgesia elicited from one supraspinal site is blocked by the prior administration of general or selective antagonists into a second site using full dose-response and time-response curves across multiple nociceptive tests. Both site specificity of antagonist effects determined by administration of antagonist into control placements dorsal or lateral to the intended sites and examination of potential antagonist-mediated hyperalgesia in a given site serve as additional important controls. Finally, agonist-induced specificity can be assessed by administering more than one opioid agonist into a given site. These approaches have been used in the study of the role of serotonergic, opioid, excitatory amino acid (EAA), cholinergic, and neurotensinergic receptors in the RVM as they pertain to the analgesic actions of opioid agonists in the vlPAG, as well as in the study of the role of opioid receptors in the vlPAG as they pertain to the analgesic actions of opioid agonists in the amygdala.

### a. Opioid Analgesia in the vlPAG and Serotonin Antagonists in the RVM

As indicated, one direct projection between the vlPAG and the NRM contains serotonin in 55–63 percent of the fibers (Beitz, 1982b), and au-

A

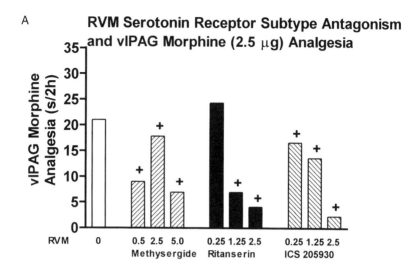

B

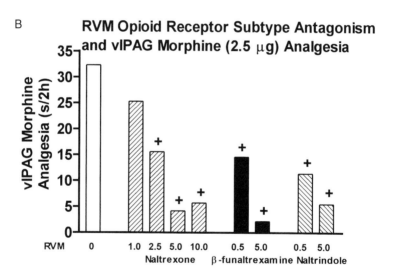

**Fig. 17.** (*A*) Administration of general (methysergide, μg), 5-HT$_2$ (ritanserin, μg), or 5-HT$_3$ (ICS205930, μg) serotonin receptor subtype antagonists into the RVM significantly (crosses display significant effects) and dose dependently reduces morphine analgesia elicited from the vlPAG on the tail-flick test, thereby implicating these serotonin receptor subtypes in the RVM in the modulation of this analgesic response through a vlPAG-RVM circuit. Administration of antagonist into sites dorsal and lateral to the RVM failed to exert effects, indicating anatomical specificity of action (Kiefel and Bodnar, 1992; Kiefel, Cooper, and Bodnar, 1992). (*B*) Administration of general (naltrexone, μg), μ (β-FNA, μg), or δ (NTI, μg) serotonin receptor subtype antagonists into the RVM significantly and dose dependently reduces mor-

toradiography confirmed the presence of 5-HT$_2$ and 5-HT$_3$ serotonin receptors on RVM neurons (Pazos, Cortes, and Palacios, 1985; Waeber et al., 1988), which have been implicated in analgesic processes (Paul and Phillips, 1986; Hasegawa et al., 1990). RVM serotonin or serotonergic agonists produce analgesia (Llewelyn, Azami, and Roberts, 1983, 1984), and the general serotonergic antagonist, methysergide, in the RVM blocks vlPAG SPA (Aimone and Gebhart, 1986). Moreover, RVM lidocaine blocks vlPAG morphine analgesia (Proudfit, 1980; Gebhart et al., 1983b; Urban and Smith, 1994a). Our laboratory (Kiefel, Cooper, and Bodnar, 1992) examined the relationship between RVM serotonin and vlPAG morphine analgesia by determining whether pretreatment of either general (methysergide), 5-HT$_2$ (ritanserin), or 5-HT$_3$ (ICS205930) serotonin receptor antagonists into the RVM altered vlPAG morphine analgesia. We found that this response was significantly reduced by RVM pretreatment with either methysergide, ritanserin, or ICS205930 (Fig. 17). Basal nociceptive thresholds were not altered by these RVM serotonergic antagonists, and they failed to alter vlPAG morphine analgesia following injection into misplaced medullary sites lateral or dorsal to the RVM.

In analyzing potential relationships between serotonin immunoreactivity and RVM physiological cell types, most RVM serotonin immunoreactivity was found in NEUTRAL cells (Potrebic, Fields, and Mason, 1994; Potrebic, Mason, and Fields, 1995), whose firing rates fail to be affected by administration of opioids (Fields, Heinricher, and Mason, 1991) and whose role in pain modulation remains enigmatic (Gao, Kim, and Mason 1997; Gao et al., 1998; Mason, 1997). Thus, RVM serotonin cells appear to have a slow, steady discharge, suggesting tonic rather than phasic modulation of spinal processes, and failed to respond during vl-PAG SPA or systemic morphine analgesia. Hence, RVM serotonin cells do not appear to be integral in the mediation of supraspinal opioid analgesia, yet the antagonist data indicate that blockade of RVM serotonergic receptors prevents the full expression of vlPAG morphine analgesia. However, cells possessing serotonin receptors can be distinct from those displaying serotonin immunoreactivity. Indeed, iontophoretic applica-

---

phine analgesia elicited from the vlPAG on the tail-flick test, thereby implicating these opioid receptor subtypes in the RVM in the modulation of this analgesic response through a vlPAG-RVM circuit. Administration of antagonist into sites dorsal and lateral to the RVM failed to exert effects, again indicating anatomical specificity of action (Kiefel, Rossi, and Bodnar, 1993). +, Significant reductions in morphine analgesia elicited from the vlPAG by the serotonergic and opioid antagonists in the RVM.

tion of RVM serotonin facilitates activity of all three physiologically described classes of neurons (Roychowdhury and Heinricher, 1997). Pharmacological studies argue that medullospinal serotonergic outflow is important for descending inhibition of nociception (see reviews in Akil et al. [1984]; Yaksh [1984a, 1984b]) as well as modulation of pain facilitation (Zhuo and Gebhart, 1990, 1991, 1992). It is therefore possible that spinal serotonin "gates" modulatory effects of both descending inhibition and descending facilitation. Nevertheless, a functional role for serotonergic receptor subtypes in the RVM has been established as an important link in the successful transmission of analgesic mechanisms by vlPAG morphine.

### b. Opioid Analgesia in the vlPAG and Opioid Antagonists in the RVM

A second direct projection between the vlPAG and the NRM contains enkephalins (Beitz, 1982a), which also serve as intrinsic interneurons in the RVM (Lewis, Khachaturian, and Watson, 1985), and both μ and δ opioid receptors are present in the RVM (Atweh and Kuhar, 1977a, 1977b; Goodman et al., 1980; Moskowitz and Goodman, 1985). Kiefel, Rossi, and Bodnar (1993) examined whether RVM pretreatment of either general (naltrexone), μ (β-FNA), or δ (NTI) opioid receptor antagonists altered vlPAG morphine analgesia and found significant and dose-dependent reductions following RVM naltrexone, β-FNA, and NTI (Fig. 17). Basal nociceptive thresholds were not altered by these RVM opioid antagonists, and they failed to alter vlPAG morphine analgesia following injection into misplaced medullary sites lateral or dorsal to the RVM. Other neuroanatomical loops mediating vlPAG morphine analgesia have been identified such that administration of naloxone into the habenula significantly reduced vlPAG morphine analgesia, and administration of naloxone into the Nacc significantly reduced morphine analgesia elicited from the habenula (Ma, Shi, and Han, 1992).

Endogenous opioids in the RVM also appear to be responsible for other analgesic actions in the vlPAG because analgesia elicited by the $GABA_A$ receptor antagonist, bicuculline, in the vlPAG was significantly reduced by RVM pretreatment with either general (naltrexone) or μ-selective (CTOP) opioid antagonists (Roychowdhury and Fields, 1996). More important, bicuculline analgesia was unaffected by antagonists injected into misplaced medullary placements. Moreover, vlPAG administration of either morphine or bicuculline significantly reduced RVM on-cell firing, which was reversed by iontophoretic RVM application of naloxone (Pan and Fields, 1996), indicating that this specific electro-physiological effect acts through an opioid synapse. By contrast, vlPAG bicuculline increased RVM off-cell firing that was unaffected by nalox-one, and thereby apparently not mediated through an opioid synapse. Furthermore, RVM DAMGO depressed on-cells irrespective of analgesic activity, but only activated off-cells when analgesia was present (Hein-richer et al., 1994). Thus, available functional and physiological data in-dicate that analgesic activation of the vlPAG releases endogenous opioid peptides in the RVM that act in at least two different ways: inhibition of on-cells for bicuculline-induced analgesia, and activation of off-cells for opioid analgesia.

### c. Opioid Analgesia in the vlPAG and Excitatory Amino Acid Antagonists in the RVM

Pain-modulating neurons in the vlPAG-RVM axis contain glutamate and aspartate (Clements et al., 1987), and those vlPAG neurons project to the RVM (Wiklund et al., 1988; Beitz, 1990). L-Glutamate or NMDA elicits analgesia following vlPAG (Behbehani and Fields, 1979; Jacquet, 1988; Siegfried and Nunes de Souza, 1989) or RVM (Satoh, Oku, and Akaike, 1983; Jensen and Yaksh, 1984a, 1984b; vanPraag and Frenk, 1990; Mc-Gowan and Hammond, 1993a, 1993b) administration. RVM EEA recep-tor antagonists increase the intensity necessary for vlPAG SPA (Aimone and Gebhart, 1986). Further, general RVM EAA antagonists significantly reduce vlPAG morphine analgesia (vanPraag and Frenk, 1990). Our lab-oratory (Spinella, Cooper, and Bodnar, 1996) examined whether RVM pretreatment of either competitive NMDA (AP-7), noncompetitive NMDA (MK-801), or kainate/AMPA (CNQX) EAA receptor antagonists altered vlPAG morphine analgesia and found significant and dose-de-pendent reductions by both competitive and noncompetitive NMDA an-tagonists, but not with kainate/AMPA antagonism (Fig. 18). Microinjec-tions of the NMDA antagonists failed to alter basal nociceptive thresholds in the RVM and also failed to alter vlPAG morphine analge-sia in misplaced medullary sites. This latter control is important given the ability of NMDA antagonists to potentially exert their effects by diffusing into structures outside the injection area (Nasstrom et al., 1993; Nasstrom, Karlsson, and Berge, 1993).

The functional effects of RVM NMDA antagonists on vlPAG morphine analgesia may be related to physiologically identified pain-inhibiting and pain-facilitating off-and on-cell classes. Heinricher and McGaraughty (1998) indicate that although spontaneous firing of off-cells appears not

A

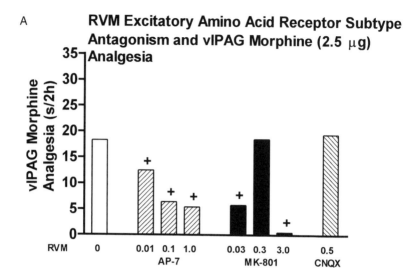

B

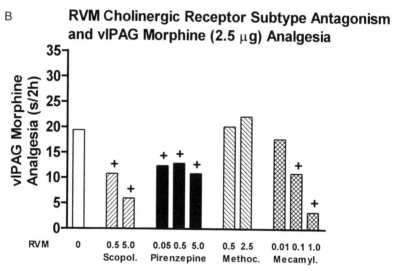

**Fig. 18.**    (*A*) Administration of competitive NMDA (AP-7, μg), noncompetitive NMDA (MK-801, μg), or AMPA/kainate (CNQX, μg) EAA receptor subtype antagonists into the RVM significantly and dose dependently reduces morphine analgesia elicited from the vlPAG on the tail-flick test, thereby implicating these NMDA receptor subtypes in the RVM in the modulation of this analgesic response through a vlPAG-RVM circuit. Administration of antagonist into sites dorsal and lateral to the RVM failed to exert effects, indicating anatomical specificity of action (Spinella, Cooper, and Bodnar, 1996). (*B*) Administration of general muscarinic (scopolamine [Scopol.], μg), $M_1$ (pirenzepine, μg), $M_2$ (methoctramine [Methoc.], μg), or general nicotinic

to be maintained by an EAA-mediated input, opioid activation of these neurons requires an NMDA-mediated excitatory process. Thus, opioid activation of an NMDA-mediated excitatory input to off-cells would account for the aforementioned behavioral effects. Moreover, although the mechanisms by which opioids activate RVM pain-inhibiting output neurons have focused classically on disinhibition, these recent studies suggest an NMDA-mediated excitatory process, presumably part of a positive feedback loop linking the PAG, RVM, and other nodes in this opioid-activated modulating system that is critical to the behavioral analgesic effects of opioids.

In addition, but probably unrelated to the aforementioned model of pain inhibition, EAA transmission appears to be responsible for RVM on-cell nociceptive activation because iontophoretic application of the EAA antagonist, kynurenate, blocked this response (Heinricher and Roychowdhury, 1997). Blockade of the on-cell burst does not affect off-cell firing, suggesting that off-cells are inhibited by neurons outside the RVM (Heinricher and McGaraughty, 1998). However, RVM administration of kynurenate before systemic morphine blocked the opioid-induced activation of off-cells and significantly reduced analgesia (Heinricher and Roychowdhury, 1997), suggesting that excitatory inputs to off-cells are critical for supraspinal opioid analgesia.

Since morphine and the opioid peptide, β-endorphin, appear to use different mechanisms of action in producing opioid analgesia (see review in Tseng [1995]), our laboratory (Spinella et al., 1999) examined whether agonist-induced specificity occurred for RVM NMDA-induced mediation of vlPAG opioid analgesia. β-Endorphin analgesia elicited from the vlPAG failed to be altered by either competitive or noncompetitive RVM NMDA antagonists at 100-fold higher doses than those blocking vlPAG morphine analgesia (Fig. 19). Therefore, these data provide evidence for opioid agonist–induced specificity within the vlPAG in the use of RVM neural circuitry to mediate specific opioid analgesic responses. Thus, it appears that RVM glutaminergic activity is an important

---

(mecamylamine [Mecamglo.], μg) cholinergic receptor subtype antagonists into the RVM significantly and dose dependently reduces morphine analgesia elicited from the vlPAG on the tail-flick test, thereby implicating these cholinergic receptor subtypes in the RVM in the modulation of this analgesic response through a vlPAG-RVM circuit. Administration of antagonist into sites dorsal and lateral to the RVM also exerted effects, indicating a wider area of medullary action (Spinella et al., 1999). +, Significantly different from control.

A

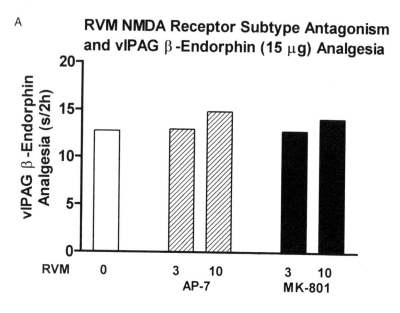

B

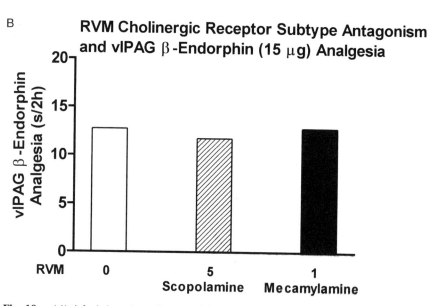

**Fig. 19.** (*A*) Administration of competitive NMDA (AP-7, μg) and noncompetitive NMDA (MK-801, μg) EAA receptor subtype antagonists into the RVM fails to alter β-endorphin analgesia elicited from the vlPAG on the tail-flick test, thereby indicating dissociations in the mediation of morphine and β-endorphin analgesia in the vlPAG. (*B*) Administration of muscarinic (scopolamine, μg) or nicotinic (mecamylamine, μg) cholinergic receptor subtype antagonists into the RVM fails to alter

determinant of the successful activation of descending pain-inhibitory systems by vlPAG morphine, but not by β-endorphin.

### d. Opioid Analgesia in the vlPAG and Cholinergic Antagonists in the RVM

Both the cholinergic agonist, carbachol and nicotine produce analgesia following vlPAG and RVM microinjections as well as microinjections into cholinergic-containing neurons in the pedunculopontine tegmental nucleus (Brodie and Proudfit, 1984, 1986; Iwamoto, 1989, 1991; Klamt and Prado, 1991). Central cholinergic agonist-induced analgesia is blocked by both muscarinic and nicotinic antagonists (Brodie and Proudfit, 1984, 1986; Iwamoto, 1989, 1991; Iwamoto and Marion, 1993a, 1993b; Guimares and Prado, 1994). The RVM has both muscarinic and nicotinic receptor subtypes (Nonaka and Moroji, 1984; London, Walter, and Wamsley, 1985; Cortes and Palacios, 1986) and receives cholinergic input from the pontomesencephalotegmental complex (Mitani et al., 1988; Rye et al., 1988). In our laboratory (Spinella, Schaefer, and Bodnar, 1997) we examined whether RVM pretreatment of either muscarinic (scopolamine), $M_1$ (pirenzepine), $M_2$ (methoctramine), or nicotinic (mecamylamine) cholinergic receptor subtype antagonists altered vlPAG morphine analgesia. We found significant and dose-dependent reductions with scopolamine, a selective effect since microinjections of scopolamine failed to alter basal nociceptive thresholds in the RVM and also failed to alter morphine analgesia elicited from the vlPAG in misplaced medullary sites (Fig. 18). However, neither $M_1$ nor $M_2$ RVM antagonism altered vlPAG morphine analgesia on the tail-flick test, and the reductions in vlPAG morphine analgesia on the jump test were smaller following pirenzepine and methoctramine. These latter effects were also not site specific, because these antagonists were effective in medullary placements dorsal and lateral to the RVM. The significant reductions in vlPAG morphine analgesia on the tail-flick and jump tests by RVM pretreatment with mecamylamine were also not site specific, because dorsal and lateral medullary placements were effective in reducing morphine analgesia as well.

Our laboratory (Spinella et al., 1999) also determined (Fig. 19) that these cholinergic antagonist effects were selective to the opioid agonist

---

β-endorphin analgesia elicited from the vlPAG on the tail-flick test, thereby indicating further dissociations in the mediation of morphine and β-endorphin analgesia in the vlPAG (Spinella et al., 1999).

employed since neither RVM scopolamine nor mecamylamine pretreatment significantly altered vlPAG β-endorphin analgesia. Although our discussions have suggested possible descending mechanisms mediating supraspinal morphine analgesia, they do not clarify a similar candidate system for supraspinal β-endorphin analgesia. Proudfit and colleagues demonstrated that met-enkephalin neurons located near the A7 noradrenergic cell group (Holden and Proudfit, 1998) innervate the dorsal horn, and these neurons can produce analgesia following local microinjection of substance P (Proudfit and Yeomans, 1995). Further, vlPAG neurons (Bajac and Proudfit, 1999) directly project to these A7 neurons that include both spinally projecting noradrenergic and enkephalin neurons. Thus, in this model, vlPAG microinjections of morphine would activate descending noradrenergic neurons, whereas vlPAG microinjections of β-endorphin would activate descending enkephalin neurons.

As indicated earlier, the RVM sends projections to such noradrenergic brain stem nuclei as the A6 and A7 cell groups, which provide major noradrenergic spinal input (Proudfit, 1988, 1992). Therefore, analgesia elicited by cholinergic agonists, glutamate, or electrical stimulation in the RVM is blocked by intrathecal α-adrenergic antagonists (Jensen and Yaksh, 1984a, 1984b; Brodie and Proudfit, 1986; Aimone, Jones, and Gebhart, 1987; Zhuo and Gebhart, 1990). The A7 cell group supports analgesia elicited by electrical stimulation and substance P (Yeomans and Proudfit, 1992; Yeomans et al., 1992), and neuroanatomical connections between the RVM and A7 cell group contain substance P and enkephalins (Holden and Proudift, 1998; Proudfit and Monsen, 1999). This RVM-A7 connection may mediate cholinergic analgesia since RVM carbachol analgesia is blocked by inactivation of the A7 cell group by either tetracaine or cobalt chloride (Nuseir, Heidenreich, and Proudfit, 1999). vlPAG morphine analgesia is differentially sensitive to spinal pharmacological manipulations as a function of the locus of the nociceptive stimulus in that this response is reduced by intrathecal $\alpha_2$-adrenoceptor, serotonergic, and muscarinic antagonists on the tail-flick test but is reversed by intrathecal muscarinic antagonists, but not by intrathecal $\alpha_2$-adrenoceptor, serotonergic, or opioid antagonists on a foot-withdrawal response (Fang and Proudfit, 1996). Furthermore, intrathecal $\alpha_1$ adrenoceptor antagonists potentiated vlPAG morphine analgesia on the foot-withdrawal response but reduced vlPAG morphine analgesia on the tail-flick test, indicating respective pronociceptive and antinociceptive actions of spinal $\alpha_1$-adrenoceptors in central morphine analgesia (Fang and Proudfit, 1998). A different pattern of responses occurs following microinjection of morphine into the A7 noradrenergic cell group such that one population of cells facilitates nociception through spinal $\alpha_1$-adrenoceptor activation while a second population of cells blocks nociception through spinal $\alpha_2$-adrenoceptor activation (Holden, Schwartz, and Proudfit, 1999). Indeed, blockade of $GABA_A$ receptors with bicuculline in the A7 noradrenergic cell group produces analgesia that is blocked by spinal $\alpha_2$-, but not by $\alpha_1$-adrenoceptor antagonists (Nuseir and Proudfit, 2000).

Although this series of studies indicated that the integrity of vlPAG morphine, but not β-endorphin, analgesia is dependent on serotonergic, opioid, NMDA, and to a less-specific extent, cholinergic receptor subtypes in the RVM, Urban and Smith (1993, 1994b) have demonstrated that RVM neurotensin modulates vlPAG morphine analgesia in an opposite manner such that RVM neurotensin significantly reduces vlPAG morphine analgesia and RVM neurotensin antagonists significantly enhance this response.

### e. Opioid Analgesia in the Amygdala and Opioid Antagonists in the vlPAG

Opioid-opioid interactions between sites have been described between the vlPAG and RVM, between the vlPAG and habenula, and between the habenula and Nacc in which opioid agonists in one site are blocked by opioid antagonists in a second site (Ma, Shi, and Han, 1992; Kiefel, Rossi, and Bodnar, 1993). The amygdala supports analgesia elicited by either morphine (Rodgers, 1977, 1978; Helmstetter, Bellgowan, and Tershner, 1993), an enkephalinase inhibitor (Al-Rodhan, Chipkin, and Yaksh, 1990), neurotensin (Kalivas et al., 1982), or carbachol (Klamt and Prado, 1991). Further amygdala lesions reduce systemic morphine analgesia on thermal and inflammatory tests (Manning and Mayer, 1995a, 1995b). Opioid analgesia elicited from the amygdala appears to be mediated by μ but not δ or κ receptors on the tail-flick test (Helmstetter, Bellgowan, and Poore, 1995; Helmstetter et al., 1998), but by both μ and κ receptors on the jump test (Pavlovic, Cooper, and Bodnar, 1996b; Pavlovic and Bodnar, 1998a). The latter studies examined the relationship between opioid analgesia elicited from the amygdala and opioid synapses in the vlPAG and found that both amygdala morphine and β-endorphin analgesia were significantly reduced by general and $\delta_2$, but not by μ opioid antagonists in the vlPAG. In contrast to the inability of the $\kappa_1$ agonist U50488H to produce analgesia in either the vlPAG, LC, or RVM (Bodnar, Paul, and Pasternak, 1991; Rossi, Pasternak, and Bodnar, 1994),

U50488H analgesia on the jump test in the amygdala was significantly reduced by $\kappa_1$ antagonist pretreatment in the amygdala, and by either general, $\mu$, or $\delta_2$ antagonist pretreatment in the vlPAG (Pavlovic and Bodnar, 1998a). That an opioid synapse in the vlPAG is necessary for the full expression of amygdala opioid analgesia was also supported by DAMGO analgesia in the basolateral amygdala reduced by general and $\mu$, but not by $\beta$-endorphin$_{1-27}$, in the vlPAG (Tershner and Helmstetter, 1995). Furthermore, vlPAG and RVM lidocaine injections blocked DAMGO analgesia elicited from the basolateral amygdala (Helmstetter et al., 1998).

Thus, the pharmacological technique by which selective antagonists are applied to one pain-inhibitory site that is purportedly downstream from a second pain-inhibitory site was successful in identifying a potential circuit (or series of circuits) through which this descending pain inhibition takes place. This technique does not address whether one site is necessary and sufficient to produce pain inhibition independent of the ability of a second site to produce pain inhibition. If so, one would then not expect to observe interactions between independent sites. Alternatively, and more appropriately for systems theorists, multiple opioid pain-inhibitory sites work in some interactive way such that full or partial activation of one part of the system would synergize with the full or partial activation of another part of the system. The following section summarizes the results of such studies indicating that opioids act to produce pain inhibition through a multiple activation of supraspinal and spinal sites of action.

### 9. Interactions between Supraspinal Opioid Analgesia Sites: Synergy Studies

Since the initial seminal studies by Yeung and Rudy (1980), the presence of synergistic analgesic interactions has been used to assess functional relationships between spinal and supraspinal opioid systems (e.g., Roerig et al., 1984; Roerig, Fujimoto, and Tseng, 1988; Roerig and Fujimoto, 1989; Siuciak and Advokat, 1989; Miyamoto et al., 1991; Pick et al., 1992; Pick, Nejat, and Pasternak, 1993). Our laboratory (Rossi, Pasternak, and Bodnar, 1993, 1994) examined whether synergistic analgesic interactions for morphine and selective opioid agonists occurred between pairs of supraspinal sites (Fig. 20). First, we determined whether functional interactions occurred between vlPAG and RVM, between vlPAG and LC, and between RVM and LC. Full analgesic dose-response curves for each agonist were ascertained in each site alone, and subsequently in pairs of sites in which one site would receive a fixed, subthreshold agonist dose, and the second site would receive a range of subthreshold doses. Thus, a dose of morphine that failed to increase tail-flick latencies when administered into either the vlPAG or RVM alone produced a significant and marked analgesia following simultaneous administration. This analgesia was more marked for vlPAG-RVM interactions than for either LC-RVM or vlPAG-LC interactions. Indeed, administration of a fixed subthreshold dose of morphine into the vlPAG produced a 10-fold leftward shift in the morphine dose-response curve in the RVM, while the subthreshold dose of morphine in the RVM produced a 3-fold leftward shift in the morphine dose-response curve in the vlPAG.

Second, we determined the opioid receptor subtypes involved in vlPAG-RVM interactions by observing that simultaneous administration of subthreshold doses of the $\mu$-selective agonist, DAMGO, into both sites produced a synergistic interaction as did simultaneous administration of subthreshold doses of the $\delta_2$-selective agonist, DELT (Fig. 20). Indeed, if a subthreshold dose of a $\mu$ agonist was applied to one site, and a subthreshold dose of a $\delta_2$ agonist was applied to the second site, synergistic analgesic interactions also occurred, implying that these interactions involve pathways rather than receptors per se. Fields and coworkers (Harasawa, Fields, and Meng, 2000) possibly indicate a potential mechanism of action for the $\delta_2$-mediated effects. Their studies showed that the analgesic effects of RVM DELT are accompanied by an inhibition of RVM on-cell pain-facilitatory effects, and a shortening of the pause in RVM off-cell pain-inhibitory effects, consistent with reports for $\mu$-selective agonists (Fields, Heinricher, and Mason, 1991; Heinricher, Morgan, and Fields, 1992; Heinricher et al., 1994), thus forming a possible basis for synergy between $\mu$ and $\delta_2$ agonists both within the RVM and between the vlPAG and RVM. By contrast, simultaneous administration of either $\delta_1$ or $\kappa_1$ opioid agonists failed either to elicit analgesia in single sites or to produce synergistic interactions with $\mu$ agonists when applied simultaneously to pairs of sites.

Pavlovic and Bodnar (1998a) examined whether synergistic analgesic interactions occurred between the amygdala and vlPAG for morphine and $\beta$-endorphin. Simultaneous administration of subthreshold doses of morphine into both sites produced a synergistic interaction, as did simultaneous administration of subthreshold doses of $\beta$-endorphin. This antinociceptive interaction persisted when subthreshold doses of

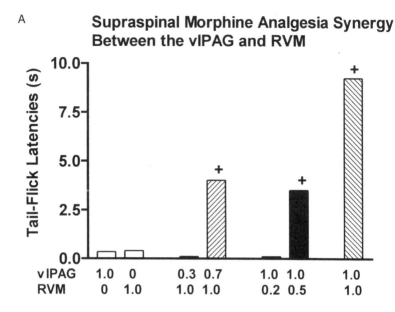

A

## Supraspinal Morphine Analgesia Synergy Between the vlPAG and RVM

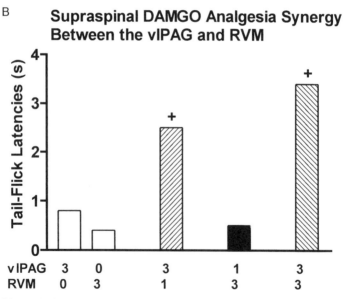

B

## Supraspinal DAMGO Analgesia Synergy Between the vlPAG and RVM

**Fig. 20.** (*A*) Analgesic synergistic multiplicative interactions occur between the vlPAG and the RVM for morphine analgesia. Whereas low subthreshold doses (1 μg) administered into each site alone failed to alter latencies, simultaneous administration of morphine into the RVM and vlPAG produced marked (~8- to 10-fold)

β-endorphin in the amygdala were coadministered with subthreshold doses of morphine in the vlPAG. However, no interaction was observed following coadministration of subthreshold doses of morphine into the amygdala and β-endorphin into the vlPAG, presumably because β-endorphin activates a different neurochemical circuit within the vlPAG relative to morphine (see review in Tseng [1995]).

This section has summarized the identification and localization of opioid peptides and opioid receptors using pharmacological, biochemical, and molecular evidence; the role of these peptides, receptors, and exogenous delivery of opiate drugs in the activation of identified, descending supraspinal pain-inhibitory systems using pharmacological, behavioral, molecular, and electrophysiological evidence; and, finally, specific lines of pharmacological, behavioral, and electrophysiological research that enable us to study relationships between specific supraspinal sites in mediating opioid analgesic responses. Although these data provide insight into how these opioid systems work to inhibit pain, they do not identify the naturalistic circumstances that activate endogenous pain inhibition. Coincident with the discovery of analgesic responses elicited by opiate microinjections and electrical stimulation of specific brain sites, and coincident with the discovery of opiate receptors and endogenous opioid peptides, came the observation that acute exposure to stressful situations can evoke analgesic responses. Such coincident discoveries gave us the opportunity to link systematically the evaluation of the pharmacological and physiological substrates of "stress-induced analgesia" with that emerging from the study of opioid mechanisms of analgesia. The next section summarizes these strategies.

---

increases in latencies, indicating the presence of synergy. Holding the subthreshold dose in one site constant and systematically lowering the dose in the other site demonstrated dose-dependent synergistic effects. Analgesic synergistic interactions were also observed for morphine for vlPAG and LC as well as for LC and RVM combinations. (From Rossi, Pasternak, and Bodnar, 1993). (*B*) Synergistic multiplicative analgesic interactions are observed following simultaneous administration of subthreshold doses of the μ-selective agonist, DAMGO. Whereas low subthreshold doses (3 μg) administered into each site alone failed to alter latencies, simultaneous administration of DAMGO into the RVM and vlPAG produced marked (~3-fold) and dose-dependent increases in latencies, indicating the presence of synergy. Analgesic synergistic interactions also were observed for the δ-selective agonist, deltorphin, for vlPAG and RVM combinations and for deltorphin and DAMGO for vlPAG and RVM combinations. +, Significantly different from control. (From Rossi et al., 1994.)

## B. Analgesia Induced by Stress
## and Environmental Variables

The evolutionary significance of intrinsic CNS pathways dedicated to pain inhibition depends on the importance of dampening an organism's normal reaction to painful stimuli during times of stress. Thus, reducing pain responses permits normal mating behavior under difficult circumstances, especially for the female, and such analgesic activation is one important component of overall organismic responses to fear, anxiety, and aggression (Bolles and Fanselow, 1980; Fanselow, 1986; Bandler and Shipley, 1994; Bandler and Keay, 1996; Bandler, Price, and Keay, 2000; Fendt and Fanselow, 2000). The wide variety of stressors that can induce analgesia is not to be underestimated nor is the inclusion of the large number of neuronal structures mediating brain and behavioral arousal (chapter II).

Early studies designed to examine stress-induced analgesia were conducted soon after the discovery of the endogenous opioid peptides. Initial work found that whereas acute exposure to inescapable foot shock increased tail-flick latencies and inhibited stereospecific central opioid binding, daily repeated exposure to inescapable foot shock failed to elicit analgesia, suggesting that the acute stress analgesia adapted with repeated exposure (Akil et al., 1976; Chance et al., 1977, 1978; Madden et al., 1977; Hayes et al., 1978). Similar patterns of analgesia and adaptation occurred following exposure to cold-water swims, glucoprivation, food deprivation, and immobilization (for reviews see Amir, Brown, and Amit [1980]; Bodnar et al. [1980b] Chance [1980]).

### 1. An Opioid Role

Initial studies expecting to reveal a prominent role of the endogenous opioids in stress-induced analgesia yielded surprising results. Hence, high doses of naloxone only partially antagonized certain forms of foot-shock analgesia (e.g., Madden et al., 1977; Jackson, Coon, and Maier, 1979), continuous cold-water swim (CCWS) analgesia (Bodnar et al., 1978b), and conditioned analgesia (Chance, 1980), and each failed to display analgesic cross-tolerance with morphine (Bodnar et al., 1978c). Although naloxone blocked analgesia induced by either immobilization or food deprivation (Amir, Brown, and Amit, 1980; McGivern and Berntson, 1980), glucoprivic analgesia displayed cross-tolerance and synergy with morphine analgesia but was unaffected by naloxone (Bodnar, Kelly, and Glusman, 1979; Spiaggia et al., 1979) (Table 7). Subsequent studies indicated that the parameters of the stressor predicted in part its opioid or nonopioid effects (Table 7). Thus, prolonged, intermittent foot shocks produced analgesia blocked by naloxone and cross-tolerant with morphine, and brief, continuous foot shocks produced analgesia insensitive to morphine cross-tolerance or naloxone reversal (Lewis, Cannon, and Liebeskind, 19890; Lewis, Sherman, and Liebeskind, 1981). Moreover, exposure to 80 but not 20 shocks elicited a naltrexone-sensitive analgesia (Grau et al., 1981).

The opioid-mediated forms of shock analgesia appeared related to a loss of controllability over the situation and thus were linked conceptually to learned helplessness (Maier et al., 1983). Indeed, a brief reexposure to inescapable shock 24 h after initial, prolonged, intermittent shock exposure produced analgesia and learned helplessness, which were both naloxone reversible (Jackson, Coon, and Maier, 1979; Maier et al., 1980), and cross-tolerant with morphine (Drugan et al., 1981). Although opioid mediation of long-term analgesia and learned helplessness can be dissociated (Hyson et al., 1982; MacLennan et al., 1982; Moye et al., 1983), controllability can determine analgesic efficacy since escapable shocks yield significantly less analgesia than yoked inescapable shocks (Maier, Drugan, and Grau, 1982). Finally, foot shocks delivered to the forepaws, but not the hindpaws, produced naloxone-sensitive analgesia (Watkins and Mayer, 1982; Watkins et al., 1982a).

Parametric variations in swims also produce differential analgesic responses. Continuous swims in warm but not cold water temperatures produced opioid-mediated analgesia responses across an array of paradigms (Bodnar et al., 1978b, 1978c; Christie, Chesher, and Bird, 1981; Christie, Trisdikoon, and Chesher, 1982; Bodnar and Sikorszky, 1983; O'Connor and Chipkin, 1984; Marek, Yirmiya, and Liebeskind, 1988). Swim patterns were also critical, since intermittent cold-water swim (ICWS) but not CCWS analgesia was blocked by morphine tolerance and opiate receptor antagonism (Bodnar et al., 1978b, 1978c; Girardot and Holloway, 1984a, 1984b). Both swim severity (Terman, Morgan, and Liebeskind, 1986) and genetic variables (Mogil et al., 1996d) emerged as integral in defining opioid mediation of swim-stress-induced analgesia. Indeed, opioid forms of stress-induced analgesia were cross-tolerant with opioid forms of SPA (Terman, Penner, and Liebeskind, 1985). These data suggested that it was not the stimulus per se that was producing each form of analgesia but, rather, the coding of the stimulus as a stressor, and this opioid-nonopioid dichotomy together with pituitary-adrenal modu-

**TABLE 7. Pharmacological and Physiological Profiles of Defferent Forms of Stress-Induced Analgesia**

| Sensitivity to Manipulation | Opioid-Neurohormonal *Prolonged Intermittent Foot Shock* | Opioid Neurohormonal *2-Deoxy-D-glucose Glucoprivation* | Opioid-Neural *Forepaw Shock* | Nonopioid-Neurohormonal *Continuous Cold-Water Swims* | Nonopioid-Neural *Brief Continuous Foot Shock* | Nonopioid-Neural *Hindpaw Shock* |
|---|---|---|---|---|---|---|
| A. Opioids | | | | | | |
| Naloxone reversal | Yes | No | Yes | No | No | No |
| Tolerance adaptation | Yes | Yes | Yes | Yes | No | No |
| Morphine cross-tolerance | Yes | Yes | Yes | No | No | No |
| Non-opioid cross-tolerance | No | Yes | No | N/A | N/A | N/A |
| B. Pituitary | | | | | | |
| Pituitary removal | Yes | Potentiated | No | Yes | No | No |
| Adrenal removal | Yes | ? | No | Potentiated | No | No |
| C. Physiological | | | | | | |
| Medial-basal hypothalamic lesion | Yes | Potentiated | ? | Yes | No | ? |
| Dorsolateral funiculus lesion | Yes | ? | Yes | ? | Yes | Yes |
| Nucleus raphe magnus lesion | Yes | ? | Yes | ? | Yes | Yes |
| Cholinergic blockade | Yes | Potentiated | Yes | Yes | No | No |

N/A, not applicable.

lation categorized stress-induced analgesia as opioid-neurohormonal, opioid-neural, nonopioid-neural, and nonopioid-neurohormonal (Watkins et al., 1982a).

## 2. Opioid-Neurohormonal Forms of Stress-Induced Analgesia

Opioid-mediated analgesia elicited by either prolonged, intermittent foot shock or brief reexposure to inescapable foot shock is dependent on the hypothalamo-hypophysial axis, because these responses are blocked by hypophysectomy, medial-basal hypothalamic damage, adrenalectomy, dexamethasone treatment, and adrenal demedullation, indicating involvement of both adrenocortical and sympathomedullary systems (Lewis, Cannon, and Liebeskind, 1980; Millan et al., 1980; Millan, Przewlocki, and Herz, 1980; Lewis, Sherman, and Liebeskind, 1981; Lewis et al., 1982). These forms of stress-induced analgesia are blocked by centrally acting muscarinic antagonists, transection of the dorsolateral funiculus of the spinal cord, lesions placed in the NRM, or midcollicular decerebration (MacLennan et al., 1982; Cannon et al., 1983; Lewis, Cannon, and Liebeskind, 1983; Lewis et al., 1983; MacLennan, Drugan, and Maier, 1983; Terman et al., 1984; Watkins et al., 1984a, 1984b, 1984c). The pattern of analgesic reductions by these manipulations is similar to that observed for morphine analgesia (see chapter III, section B1). However, these opioid shock-induced analgesic responses dissociate from morphine analgesia in terms of mediation by hypothalamo-hypophysial-adrenal mechanisms (Bodnar et al., 1980a; Badillo-Martinez et al., 1984a).

Glucoprivation is a second form of opioid-neurohormonal stress-induced analgesia (Bodnar et al., 1978a) (Table 7) that adapts with repeated exposure, displays cross-tolerance and synergy with morphine analgesia, but is insensitive to opioid antagonism (Bodnar, Kelly, and Glusman, 1979; Spiaggia et al., 1979). In contrast to the opioid forms of foot-shock analgesia, glucoprivic analgesia is potentiated by either medial-basal hypothalamic damage or hypophysectomy (Bodnar et al., 1979b; Badillo-Martinez et al., 1984b), and displays full reciprocal cross-tolerance with nonopioid-mediated CCWS analgesia (Spiaggia et al., 1979). Glucoprivic analgesia is potentiated by either muscarinic or dopaminergic antagonism (Bodnar and Nicotera, 1982; Sperber, Kramer, and Bodnar, 1986), but is reduced by dopaminergic agonists (Bodnar et al., 1980c). Glucoprivic analgesia displays test-specific effects following serotonergic antagonists such that 5-HT$_3$ antagonism reduces and potentiates this response on the tail-flick and jump tests, respectively (Fisher and Bodnar, 1992). By contrast, morphine analgesia is reduced on both tests by this and other serotonergic antagonists (Kiefel, Cooper, and Bodnar, 1992).

The next section describes stressors that share opioid-sensitive profiles but appear to act independently of pituitary-adrenal systems.

### 3. Opioid-Neural Forms of Stress-Induced Analgesia

Analgesia elicited by either shock delivered to the forepaws or very brief shocks to all paws displays morphine cross-tolerance and sensitivity to supraspinal and spinal opiate antagonists yet is unaffected by hypophysectomy or adrenalectomy (Watkins et al., 1982a, 1982b; Watkins and Mayer, 1982; Terman et al., 1984) (Table 7). Pontine and medullary descending control mechanisms mediate these forms of analgesia given the reductions observed following spinal dorsolateral funiculus transections, NRM lesions, but not midcollicular decerebration (Watkins et al., 1983, 1984a, 1984b; Watkins, Kinscheck, and Mayer, 1983; Terman et al., 1984). Spinal cholecystokinin systems modulate these foot-shock-induced responses since CCK-8 and its antagonists respectively reduce and potentiate foot-shock-induced analgesia (Faris et al., 1983; Watkins et al., 1985; Watkins, Kinscheck, and Mayer, 1985; Watkins et al., 1986). Food deprivation is a final form of analgesia blocked by opioid antagonists but unaffected by hypophysectomy (Bodnar et al., 1980b; McGivern and Berntson, 1980; Hamm and Lyeth, 1984; Hamm and Knisely, 1986).

To summarize, two forms of opioid-mediated foot-shock analgesia have been identified, one characterized by intermittency or long exposures and dependent on pituitary-adrenal, "higher" brain centers and bulbospinal systems for its full expression. These effects are consistent with systems involved in stimulus controllability, particularly in terms of adaptive, stress-related mechanisms associated with motivational and cognitive processes. The second is characterized by either extreme brevity or selected regional stimulation (e.g., forepaws), is dependent only on bulbospinal systems, and appears to be associated with an absence of controllability. Thus, either controllability or brevity of stressful stimuli produces opioid analgesia independent of pituitary-adrenal activation. Potential insights into the mechanistic order of opioid-neurohormonal forms of stress-induced analgesia suggest that such stressors activate the pituitary-adrenal axis, which, in turn, activate supraspinal opioid systems, whereas opioid-neural forms of stress-induced analgesia suggest that supraspinal opioid systems can be directly activated by salient though brief stimuli. One candidate system is the amygdala, which when lesioned selectively decreases opioid but not nonopioid forms of stress-induced analgesia (Helmstetter, 1992; Helmstetter and Bellgowan, 1993; Watkins, Wiertelak, and Maier, 1993; Fox and Sorenson, 1994; Pavlovic, Cooper, and Bodnar, 1996a). Indeed, a number of motivated behaviors are implicated in the mediation of opioid-neural analgesic responses to stress (Helmstetter, 1992, 1993). Such a concept is, of course, integral to the coordination of several behavioral responses to stress (e.g., sex, aggression), of which analgesia is but one component. The following two sections consider forms of stress-induced analgesia independent of opioid manipulations.

### 4. Nonopioid-Neural Forms of Stress-Induced Analgesia

Analgesia elicited by brief, continuous foot shock, a low number (20) of inescapable tail shocks, autoanalgesia, or shock delivered exclusively to the hindpaws is neither cross-tolerant with morphine analgesia nor blocked by opioid receptor antagonists (Chance, 1980; Grau et al., 1981; Lewis, Sherman, and Liebeskind, 1981; Watkins et al., 1982a, 1982b), is unaffected by hypophysectomy, adrenalectomy, or midcollicular transection, but is significantly reduced by spinal dorsal lateral funiculus transections and NRM lesions (Lewis, Sherman, and Liebeskind, 1981; Lewis et al., 1982, 1983; Watkins and Mayer, 1982; Cannon et al., 1983; Watkins et al., 1983; Watkins, Kinscheck, and Mayer, 1983) (Table 7). These forms of stress-induced analgesia are reduced by muscarinic, $\alpha_2$-adrenergic, and $H_2$ receptor antagonists (Lewis, Cannon, and Liebeskind, 1983b; Watkins et al., 1984c; Hough, Glick, and Su, 1985; Gogas et al., 1987; Gogas and Hough, 1988).

### 5. Nonopioid-Neurohormonal Forms of Stress-Induced Analgesia

The most studied form of nonopioid-neurohormonal stress-induced analgesia is elicited following CCWS (Table 7). CCWS analgesia has been classified as nonopioid because it fails to show morphine cross-tolerance, sensitivity to naloxone, reduction by enkephalinase inhibition, and potentiation following either acute $\mu_1$ antagonism or chronic naltrexone treatment, yet it is reduced by either spinal $\delta_2$ opioid antagonist pretreatment or combinations of $\mu$, $\delta$, and $\kappa$ opioid antagonists (Bodnar et al., 1978b, 1978c; Bodnar, Lattner, and Wallace, 1980; Kirchgessner, Bodnar, and Pasternak, 1982; Bodnar and Sikorszky, 1983; Yoburn et al., 1987; Vanderah et al., 1992; Watkins et al., 1992). CCWS analgesia appears neurohormonal because it is reduced by hypophysectomy, potentiated by adrenalectomy, but unaffected by adrenal demedullation and peripheral catecholamine depletion (Bodnar et al., 1979a; Marek, Ponocka, and Hartmann, 1982; Bodnar, 1984). Anterior lobe adrenocortical actions mediate CCWS analgesia through reductions following dexamethasone or medial-basal hypothalamic damage (Bodnar et al., 1980a; Badillo-Martinez, 1984b), potentiations following inhibition of corticosteroid synthesis (Mousa, Miller, and Couri, 1981, 1983), and lack of effects following removal of the posterior lobe of the pituitary gland (Kelly et al., 1993). Only hypophysectomized animals supplemented with corticosterone and L-thyroxine, to maintain normal homeostatic responses, displayed reductions in CCWS analgesia while unsupplemented hypophysectomized controls displayed a normal analgesic response (Kelly et al., 1993).

Whereas dopamine agonists or tail-pinch stress reduces CCWS analgesia, dopamine antagonists enhance this response (Bodnar et al., 1980c; Bodnar and Nicotera, 1982; Simone and Bodnar, 1982). By contrast, CCWS analgesia is potentiated by $\alpha_2$-adrenergic receptor agonists and antagonists as well as desimipramine (Bodnar, Merrigan, and Sperber, 1983; Bodnar, Mann, and Stone 1995; Kepler and Bodnar, 1988). Both muscarinic and NMDA antagonists reduce CCWS analgesia, while $H_2$ receptor antagonism potentiates this analgesic response (Sperber, Kramer, and Bodnar, 1986; Robertson, Hough, and Bodnar, 1988; Marek et al., 1991c, 1992). Whereas systemic NMDA antagonists fail to affect morphine or opioid swim-stress analgesia (Marek et al., 1991a, 1991b; Trujillo and Akil, 1991; Ben-Eliyahu et al., 1992; Vaccarino et al., 1992), RVM NMDA antagonism reduces vlPAG morphine analgesia (Spinella, Cooper, and Bodnar, 1996) and fails to alter CCWS analgesia (Hopkins et al., 1998).

Nonopioid neuropeptide systems such as thyrotropin-releasing hormone (TRH) and vasopressin are important in CCWS analgesia. Ventricular TRH potentiates CCWS and nonopioid foot-shock analgesia (Butler and Bodnar, 1984, 1987) but fails to alter morphine analgesia (Holaday and Faden, 1983; Watkins et al., 1986). Interestingly, site-specific effects of TRH within the vlPAG have been observed with anterior vlPAG TRH decreasing CCWS analgesia, posterior vlPAG TRH potentiating CCWS analgesia, and potentiations of morphine analgesia by both anterior and posterior vlPAG TRH (Robertson and Bodnar, 1993). Vasopressin-deficient Brattleboro rats fail to display CCWS analgesia while expressing normal opiate analgesia (Bodnar et al., 1980d). Vasopressin elicits analgesia that acts independently of both opioid systems and its posterior hypophyseal actions (Berntson and Berson, 1980; Berkowitz and Sherman, 1982; Kordower, Sikorszky, and Bodnar, 1982; Berson et al., 1983; Kordower and Bodnar, 1984). Hypothalamic paraventricular nucleus (PVN) lesions decrease both vasopressin and CCWS analgesia (Bodnar, 1986). PVN vasopressin projects to three areas: the posterior pituitary, the median eminence controlling anterior pituitary activity, and extrahypothalamic structures (e.g., Swanson and Kuypers, 1980). Since vasopressin and CCWS analgesia display dissociated responses following hypophysectomy (Bodnar et al., 1979a; Berson et al., 1983) and medial-basal hypothalamic damage (Bodnar et al., 1980a; Bodnar, Truesdell, and Nilaver, 1985; Badillo-Martinez et al., 1984b), it appears that these PVN vasopressin outputs do not mediate both responses, thus suggesting extrahypothalamic and possible spinal loci of action (Millan et al., 1984).

As indicated earlier, parametric variation in CWS patterns determines opioid mediation. Hence, ICWS but not CCWS analgesia is blocked by morphine tolerance and opiate antagonists (Girardot and Holloway, 1984a, 1984b). Indeed, ICWS analgesia is significantly potentiated by endopeptidase inhibitors that act on longer-chained opioids such as dynorphin A and MERGL (Chu and Orlowski, 1985; Kest, Orlowski, and Bodnar, 1991, 1992). Although inhibition of tryptophan hydroxylase with parachlorphenylalanine fails to alter CCWS analgesia (Bodnar et al., 1981), general and 5-HT$_2$ antagonists potently reduce CCWS analgesia to a greater degree than ICWS analgesia (Kiefel, Paul, and Bodnar, 1989). Surprisingly, central serotonergic mediation of morphine analgesia (Kiefel, Cooper, and Bodnar, 1992) is more similar to that observed for CCWS than for ICWS analgesia. One potential explanation is that CCWS analgesia is mediated through supraspinal serotonergic mechanisms, while ICWS analgesia is mediated through spinal 5-HT$_1$ mechanisms (Rochford and Henry, 1988). Although the RVM supported potent 5-HT antagonist-mediated reductions in vlPAG morphine analgesia (Kiefel, Cooper, and Bodnar, 1992), methysergide microinjected into the RVM potentiated CCWS analgesia and failed to alter ICWS analgesia (Hopkins et al., 1998). Although the amygdala supports both morphine (Manning and Mayer, 1995a, 1995b) and opioid-mediated stress-induced (Helmstetter, 1992, 1993; Helmstetter and Bellgowan, 1993; Fox and Sorenson, 1994; Bellgowan and Helmstetter, 1996), but not all (Watkins, Wiertelak, and Maier, 1993) analgesia, amygdala lesions failed to alter either CCWS or ICWS analgesia (Pavlovic, Cooper, and Bodnar, 1996a). Whereas inhibition of nitric oxide synthase selectively blocks $\mu$-mediated tolerance, but not analgesia (Kolesnikov, Pick, and Pasternak, 1992; Kolesnikov et al., 1993), this manipulation potently potentiates CCWS but not ICWS analgesia (Spinella and Bodnar, 1994). However, despite their differential sensitivity to opioid manipulations, both CCWS and ICWS analgesia displayed adaptation following chronic exposure, and full, reciprocal cross-tolerance (Pavlovic and Bodnar, 1993).

### 6. A Model of Collateral Inhibition for Different Forms of Analgesia

The preceding evidence suggests strongly that multiple pain-inhibitory systems exist, defined in terms of activation by either selective stressors or specific stress parameters. It is important to determine whether a selective stressor activates all or only the most appropriate analgesic system. Selye (1952) defined hierarchies of stress-adaptive responses and indicated that the most appropriate system is activated first with less appropriate systems held in reserve. If this holds true for analgesic systems, what mechanism would allow one analgesic system to be expressed while the others remain silent? Our laboratory (Kirchgessner, Bodnar, and Pasternak, 1982; Bodnar, 1984, 1986) proposed a collateral inhibition model in which selective activation of one analgesic system should actively inhibit less appropriate analgesic systems in addition to exerting its pain-inhibitory effects. Dissociations between CCWS and morphine analgesia are observed following enkephalinase inhibition, which reduces CCWS analgesia and potentiates morphine analgesia (Alleva, Castellano, and Oliverio, 1980; Bodnar, Lattner, and Wallace, 1980). Furthermore, either $\mu_1$ opioid antagonism receptors or chronic naltrexone infusion potentiates CCWS analgesia and reduces morphine analgesia (Pasternak, Childers, and Snyder, 1980; Kirchgessner, Bodnar, and Pasternak, 1982; Yoburn et al., 1987). Both medial-basal hypothalamic lesions and hypophysectomy reduce CCWS analgesia and potentiate morphine analgesia (Bodnar et al., 1979a, 1980a; Badillo-Martinez et al., 1984a, 1984b).

To test directly the collateral inhibition model, we (Steinman et al., 1990) found that coadministration of CCWS and morphine produced analgesia that failed to differ from that of CCWS itself, and that was significantly less than morphine alone (Fig. 21). Shortening the duration

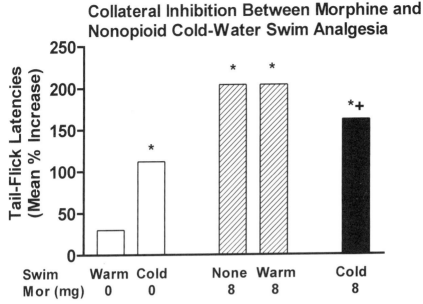

**Fig. 21.** The analgesic response following acute exposure to CCWSs (2°C for 3.5 min) has been characterized as nonopioid based on its lack of cross-tolerance with morphine analgesia and its failure to be affected by general opioid antagonist pretreatment. Pairing CWSs with morphine (Mor) (8 mg/kg) resulted in an analgesic response that was significantly lower than morphine alone despite the fact that the swims themselves produced a marked analgesia. Swimming per se did not reduce morphine analgesia because warm-water swims failed to alter morphine's analgesic response. In addition, this effect was shown to display orderly dose- and time-dependent effects, suggesting that the opioid pain-inhibitory system activated by morphine and the nonopioid pain-inhibitory system activated by CCWSs produced reciprocal collateral inhibition on each other's response in addition to producing analgesia per se. +, Significantly different from control; *, significantly different from no-swim control. (From Steinman et al., 1990.)

of CCWSs or lowering morphine doses continued to produce systematically collateral inhibition. The inhibition was not observed following temporal separation between CCWS and morphine, or when morphine was paired with warm-water swims. Thus, inhibition of opiate analgesia by prior activation of nonopioid systems appears dependent on the strength of the nonopioid manipulation, the opiate dose, and the temporal presentation of CCWS and morphine. This model was also supported by naltrexone's potentiation of CCWS and social defeat analgesia, and morphine's significant reductions of these actions (Grisel et al., 1993). Further support for collateral inhibition includes the "paradoxical" analgesic effects of naloxone and naltrexone under certain situations (Greeley et al., 1988; Westbrook and Greeley, 1990), and by the apparent nonopioid analgesic response induced by repeated saline injections followed by subsequent testing on the hot-plate test (Rochford and Dawes, 1993). These effects contrast with analgesic synergy between opioid-mediated stressors and morphine (Appelbaum and Holtzman, 1984, 1985; Sherman, Strub, and Lewis, 1984). The collateral inhibition model suggests that these multiple mechanisms of pain inhibition are activated selectively to respond to or avoid noxious situations in the most appropriate and parsimonious manner. This system of producing parallel enhancements and reductions of different forms of analgesia may allow an animal to continue responding (actively or passively) to its environment in a way that benefits the organism as a whole, and not merely as an immediate response to pain. Such a schema has been proposed recently for a range of active and passive coping responses involving identical brain circuits involved in analgesic processing (Bandler, Price, and Keny, 2000). Thus, the parameters of the stressful situation code which forms of pain inhibition might be most appropriate in terms of overall function and fit in nicely with other explanations of the phenomenon of stress-induced analgesia, such as the perceptual-defense-recuperative model (Bolles and Fanselow, 1982).

## 7. Other Environmental Forms of Analgesia

This section briefly reviews analgesia induced by vaginocervical probing (for reviews see Whipple and Komisaruk [1985]; Komisaruk and Whipple [1986]), defeat in an aggressive encounter (Miczek, Thompson, and Schuster, 1982), and pregnancy (Gintzler, 1980).

### a. Vaginocervical Probing Analgesia

Komisaruk and coworkers found that vaginocervical probing in rats and humans produce analgesia across a range of pain tests without producing anesthesia or motor dysfunction (Steinman et al., 1983; Whipple and Komisaruk, 1985; Komisaruk and Whipple, 1986). Corresponding elevations in pain thresholds and blood pressure are mediated through separate mechanisms (Catelli, Sved, and Komisaruk, 1987). Vaginocervical probing analgesia is reduced by 50 percent by spinal transection (Watkins et al., 1984b). Opioid mediation of vaginocervical probing-induced analgesia is test specific with changes in tail-flick latencies and tail shock–induced vocalization respectively sensitive and insensitive to opioid antagonism (Hill and Ayliffe, 1981; Komisaruk and Whipple, 1986). Vagino-cervical probing analgesia also interacts with nonopioid systems given its cross-tolerance with CCWS analgesia (Bodnar and Komisaruk, 1984). Whereas vaginocervical probing analgesia is potentiated by an inhibitor of neuropeptide degradation, the glycinergic antagonist, strychnine, re-duces this analgesic response (Roberts, Beyer, and Komisaruk, 1985; Heller et al., 1986). Both noradrenergic and serotonergic systems have been implicated in vaginocervical probing-induced analgesia as func-tions of perfusate as well as antagonist studies (Steinman et al., 1983). In contrast to stress-induced analgesia that has been tested in isolated in-stances in humans (Willer and Ernst, 1986), vaginocervical probing-in-duced analgesia has been studied in women, and its analgesic mecha-nisms have been defined in terms of the force applied, the regions stimulated, and the subjective relationship between pleasure and anal-gesia (see review in Whipple and Komisaruk [1985]).

## b. Defeat-Induced Analgesia

Among different models of rodent aggression, the resident-intruder par-adigm possesses a great deal of behavioral control because when an in-truder rodent is placed into the home cage of another rodent (resident), the resident typically wins despite weight and size differences. Defeat in an aggressive encounter produced analgesia with its degree increasing with the intensity of the encounter (Miczek, Thompson, and Schuster, 1982), and the victorious rodent, though often wounded, failed to dis-play analgesia (Siegfried, Frischnecht, and Waser, 1984; Miczek, Thomp-son, and Schuster, 1986; Siegfried et al., 1987). Defeat-induced analgesia is blocked by general opioid antagonists and displays tolerance and cross-tolerance with morphine (Miczek, Thompson, and Schuster, 1982; Rod-gers and Hendrie, 1983; Teskey, Kavaliers, and Hirst, 1984; Rodgers and Randall, 1985; Kulling et al., 1987; Siegfried et al., 1987). Naltrexone's reduction in defeat-induced analgesia is observed in the PAG and ar-cuate nucleus (Miczek, Thompson, and Schuster, 1985). By contrast, defeat-induced analgesia is unaffected by disruptions of the pituitary-adrenal axis (Thompson et al., 1988). Defeat-induced analgesia alters central $\mu$ opioid receptor number and $\beta$-endorphin immunoreactivity (Miczek, Thompson, and Schuster, 1986; Kulling et al., 1988). Opioid mediation depends on longer as compared with shorter exposure to de-feat, with the resident's scent a critical variable in eliciting both forms of analgesia (Rodgers and Randall, 1986a, 1986b, 1986c). Nonopioid but not opioid forms of defeat-induced analgesia are mediated by benzodiazepine (Rodgers and Randall, 1987a, 1987b, 1987c). Indeed, brief exposure to a natural predator produces nonopioid, benzodiazepine-sensitive analgesia, whereas prolonged exposure produces analgesia sensitive to both opioid and benzodiazepine systems (Kavaliers, 1988). Such data indicate the importance of parametric considerations in the activation of opioid and nonopioid systems even in "naturalistic" situations.

## c. Pregnancy-Induced Analgesia

Nociceptive thresholds increase over the latter stages of both actual pregnancy and pseudopregnancy and during parturition across several species including humans (Gintzler, 1980; Cogan and Spinato, 1986; Gintzler and Bohan, 1990; Kristal et al., 1990; Whipple, Josimovith, and Komisaruk, 1990) over a time course similar to pregnancy-induced activation of endogenous opioids (Cstonas et al., 1979; Facchinetti et al., 1982; Wardlaw and France, 1983). Opioid antagonists block pregnancy-induced analgesia in pregnant and pseudopregnant rats (Gintzler, 1980; Gintzler and Bohan, 1990). However, manipulations of the pituitary-adrenal axis fail to alter these responses (Baron and Gintzler, 1984, 1987). Spinal $\delta$ and $\kappa$ but not $\mu$ opioid receptors appear to modulate pregnancy-induced analgesia on the basis of selective antagonist and antisera studies (Sander, Portoghese, and Gintzler, 1988; Sander, Kream, and Gintzler, 1989; Dawson-Basoa and Gintzler, 1997). Combined increases of the ovarian hormones, 17$\beta$-estradiol and progesterone, increase nociceptive thresholds during pseudopregnancy and enhance $\kappa$ receptor agonist-induced analgesia by modulating spinal dynorphin levels (Dawson-Basoa and Gintzler, 1993, 1996; Medina et al, 1993). Thus, as gestation progresses, spinal dynorphin $A_{1-17}$ and dynorphin $A_{1-8}$ precursor levels decline, and peptide levels increase to crescendo at the time of parturition (Medina, Gupta, and Gintzler, 1995). The enhanced processing of dynorphin precursor intermediates may represent the initial biochemical level of adaptation of spinal dynorphin neurons to the increased demands of pregnancy. In contrast to dynorphin-induced facilitation of pregnancy-induced analgesia, OFQ/N significantly reduces analgesia induced by pregnancy or sex hormones (Dawson-Basoa and Gintzler, 1997) by inhibiting the release of spinal met-enkephalin (Gintzler et al., 1997). Intrathecal $\delta$ but not $\mu$ opioid antagonists significantly reduce normal and hormone-stimulated pregnancy analgesia (Dawson-Basoa and Gintzler, 1997), whereas $\delta$ and $\kappa$ opioid antagonists act synergistically to reduce these analgesic responses (Dawson-Basoa and Gintzler, 1998). Supraspinal analgesic systems also are recruited during normal and hormonally-stimulated pregnancy through activation of the hypogastric nerve, since resection blocks both analgesic responses (Liu and Gintzler, 1999). Because intrathecal administration of $\alpha_2$- but not $\alpha_1$-adrenergic antagonists blocks normal as well as hormonally stimulated pregnancy-induced analgesia (Liu and Gintzler, 1999), the A7 noradrenergic cell group that mediates supraspinal morphine analgesia (Proudfit and Yeomans, 1995; Fang and Proudfit, 1996; Holden and Proudfit, 1998; Holden, Schwartz, and Proudfit, 1999) may mediate pregnancy-induced analgesia also.

Thus, it appears that pregnancy-induced analgesia uses both spinal and supraspinal opioid-mediated systems, and pregnancy per se is not necessary to trigger this response. Since those gonadal hormones altered during pregnancy produce analgesia in the absence of physiological pregnancy, those signals appear integral in driving the interaction between analgesic systems and the ultimate successful output of reproductive behavior, pregnancy, and successful delivery of the offspring. We now turn to a more detailed evaluation of the interactions between sex steroids and endogenous opioid function, especially by focusing on the enkephalin gene.

## C. Hormonal Control of the Enkephalin Gene: One Paradox and Three Solutions

The effects of sex hormones on enkephalin RNA show that we can turn on a gene at a specific place in the nervous system at a particular time and that the resultant gene product has behavioral consequences. This is easily understood from the molecular biological facts and the behavioral studies quoted below. However, the direction of change of expected synaptic activity in VMH neurons presented a paradox that needed a lot of work for its solution.

The administration of estrogen turns on the enkephalin gene within specific neurons in the hypothalamus (Fig. 22) (Romano et al., 1988, 1989; Priest, Eckersell, and Micevych, 1995; Priest, Borsook, and Pfaff, 1996), in females much better than in males (Romano et al., 1990). Stimulation of the transcription of enkephalin mRNA is well correlated with the occurrence of lordosis (Lauber et al., 1990). This effect is transcriptional (Fig. 23) (Zhu and Pfaff, 1994, 1995, 1998; Yin, Kaplitt, and Pfaff, 1995; Zhu et al., 1996; Zhu, Dellovade, and Pfaff, 1997a, 1997b; Zhu et al., 1997, 2000a, 2000b). Its functional meaning appears to include some

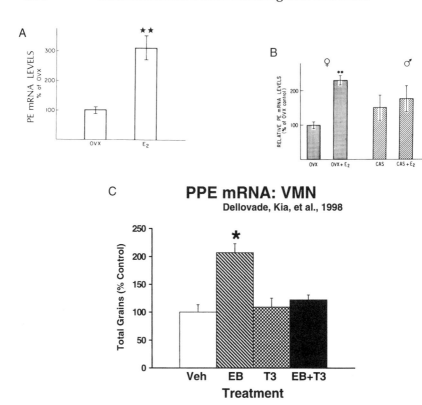

**PPE mRNA: VMN**
Dellovade, Kia, et al., 1998

**Fig. 22.** Administration of estrogen induces higher levels of enkephalin mRNA in the VMH. (*A*). Slot-blot filter hybridization assays demonstrated a 3.1-fold induction by estradiol (E₂) treatment. This discovery was replicated using in situ hybridization histochemistry, which showed a 3.3-fold induction (from Romano et al., 1988). Several features of this induction parallel lordosis behavior itself. (*B*) The induction is much stronger in genetic female rats than in genetic males. Neither estradiol (E₂) nor testosterone (T) induced higher levels of preproenkephalin (PPE) mRNA in castrated (CAS) male rats, even under the same circumstances in which the estrogenic induction was replicated in genetic females (from Romano et al., 1990). (*C*) Even as estradiol benzoate (EB)-stimulated lordosis behavior can be significantly reduced by coadministration of thyroid hormones (triiodothyronine, T₃), the estrogenic induction of PPE mRNA in the VMH can be abolished by coadministration of thyroid hormones (Dellovade et al., 1999). Veh, vehicle; PE, proenkephalin. ☆, Significant increase by CE₂; *, significant increase relative to control.

certain range of behavioral effects that eventually can facilitate lordosis (Pfaus and Pfaff, 1992; Nicot et al., 1997). The rapidity of the effect of estrogen on enkephalin transcription within 1 h (Romano et al., 1989; Zhu and Pfaff, 1994, 1995, 1998; Zhu et al. 1996; Zhu, Dellovade, and Pfaff, 1997a, 1997b; Zhu et al. 1997, 2000a, 2000b) is surprising but may be correlated with the effects of opioid peptides on lordosis when opioid-active agents are given around the time of estrogen priming (Torii, Kuba, and Sasaki, 1995, 1996, 1997). With respect to reproductive behavior, the functional meaning of the synthesis from the enkephalin gene is not likely to be limited to lordosis itself. Stress leads to an increase in opioid peptide synthesis; stress interacts with the effects of estrogen on the mouse enkephalin gene promoter (Priest, Borsook, and Pfaff, 1996) (Fig. 24), and, indeed, the application of stress seems to be essential for aspects of the phenotype of an enkephalin gene knockout (Ragnauth et al., 2001). There also may be a relationship here with nociceptive control, because enkephalin inputs have been located directly on the on-cells in RVM that are directly inhibited by μ agonists (Mason, Back, and Fields, 1992). It is not clear, however, whether such enkephalin inputs are directly under sex hormone control. Altogether, charting the interactions between the milieu of sex hormones and the level of environ-

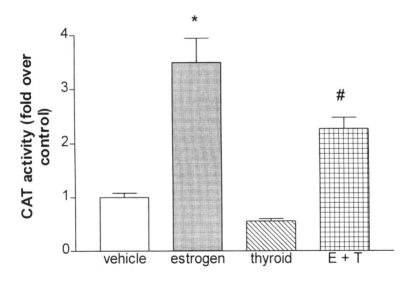

**Treatment groups**

**Fig. 23.**    The estrogenic facilitation of PPE gene expression is transcriptional. Here, using the PPE promoter and a reporter gene (CAT), the estrogenic facilitation of enkephalin gene expression is replicated and extended. In addition, at the transcriptional level, coadministration of estrogenic and thyroid hormones (E+T) significantly reduces the estrogenic effect. *, Significant estrogen-induced increases in CAT gene activity; #, significant reversal of the estrogen-induced increase by thyroid hormone cotreatment.

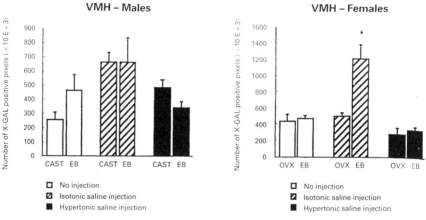

**Fig. 24.**   In the VMH of mice, the estrogenic effect (EB) on PPE mRNA depends on not only gender but also level of stress. Low stress: no injection; moderate stress: isotonic saline injection; high stress and pain: hypertonic saline injection. In our experiment here with transgenic mice studying transcription through the PPE promoter, in situ hybridization measurement of the reporter gene's expression demonstrated a significant effect only in genetic females under conditions of mild stress. CAST, castrated. *, Significant estrogen-induced increases in PPE mRNH in the ventromedial hypothalamus following isotonic saline injections relative to no-injection controls or hypertonic saline injections. (From Priest, Borsook, and Pfaff, 1996.)

mental stress will be crucial for understanding how enkephalin synthesis and deposition could play a role in protecting instinctive behaviors against the disruptive effects of stress.

Finally, genes for opioid peptide receptors also are influenced by sex hormones. In the hypothalamus, the administration of estrogen elevates levels of mRNA for the $\mu$ and $\delta$ (Quinones-Jenab et al., 1996, 1997) opioid receptors.

The enkephalin transcriptional increase caused by estrogen represents a *paradox*, among functional gene/behavior relations proven in this neuroendocrine system, because enkephalin ordinarily inhibits electrical activity, whereas increased electrical activity in these hypothalamic neurons is required for estrogen-facilitated lordosis (Kow and Pfaff, 1988). That is, the paradox is that estrogens turn on the gene for enkephalin, but enkephalins as opioid peptides would usually be expected to *inhibit* electrical activity. We ordinarily would expect an estrogen-stimulated gene product to foster female reproductive behavior, but it has been well established that *increased* electrical activity in VMH neurons is associated with estrogen-facilitated lordosis (Fig. 25). Therefore, enkephalins as in-hibitory neuropeptides would contrast dramatically with what would be expected electrophysiologically from a lordosis-facilitated gene product. How can this be possible?

Years of work have yielded three solutions to this paradox: first, enkeph-alinergic projections to the POA; second, enkephalins' ability to foster electrical activity in PAG output neurons; and third, disinhibitory mech-anisms within both the VMH and PAG.

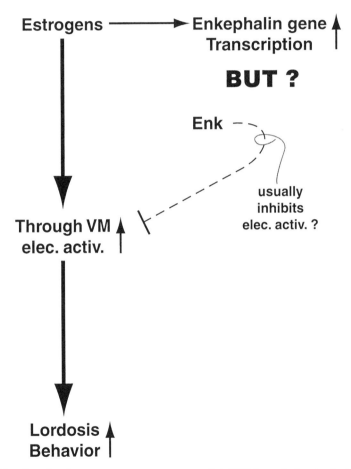

**Fig. 25.**   Paradox: Estrogens turn on ventromedial (VM) hypothalamic electrical ac-tivity, as required for the facilitation of lordosis behavior. Estrogens also turn on the PPE gene, *but* the ordinarily expected electrophysiological effect of the enkephalin re-leased would be inhibitory. Therefore, there is a mismatch among these well-elaborated sets of neurobiological data.

*First,* an important finding by Akesson and Micevych (Akesson, Simerly, and Micevych, 1988, 1991) showed that enkephalinergic neurons in the rat VMH project to the POA. Since the major force of preoptic neurons is to inhibit lordosis (Pfaff, 1980), enkephalin acting as an inhibitory neuropeptide in the POA would serve to inhibit an inhibitor, thus to release lordosis behavior. This connectivity by itself would serve to solve the paradox.

Second, VMH projections to the PAG have received a lot of attention. As one part of the general principle that estrogen-binding neurons tend to project to other estrogen-receptive areas (Pfaff and Keiner, 1973; Cottingham and Pfaff, 1986), it has been described in detail how estrogen-receiving VMH neurons project to estrogen-binding PAG areas (Turcotte and Blaustein, 1999). In line with the generalization that steroid hormone–binding areas are conserved markedly across vertebrate species, ER immunoreactive PAG neurons have been discovered in the female cat, as well (Vanderhorst et al., 1998). This replicates and extends findings in the female rat (Pfaff and Keiner, 1973). Therefore, it is likely that estrogen-binding VMH neurons have the opportunity to synapse on "output neurons" in the lateral portions of the PAG. Important, therefore, is the fact that low concentrations of opioid agonists—far from inhibiting such neurons—can actually potentiate the electrical responsiveness of such midbrain PAG neurons to NMDA agonists (Fig. 26) (Kow et al., 1998). Such a finding, important for understanding female reproductive behavior, falls into the tradition established by the publications of Gintzler (Gintzler, Chan, and Glass, 1987; Shen, Crain, and Ledeen, 1991; Dawson-Basoa and Gintzler, 1993; Wang and Gintzler, 1994, 1995; Chakrabarti et al., 1998a, 1998b). We now understand this bimodal opioid action to involve $G\beta\delta/G s\alpha$ stimulation of adenylyl cyclase activity through a PKC-mediated phosphorylation (Chakrabarti et al., 1998a, 1998b). The use of $\mu$ opioid agonists in Kow's experiments is especially telling since the $\mu$ opioid receptor is required for enkephalin-induced analgesia (Sora, Funada, and Uhl, 1997). In fact, considering the NMDA linkage in the PAG, upregulation of NMDA receptors follows administration of enkephalins (Bhargava, Kumar, and Bian, 1997), and, most amazingly, NMDA receptors are important components in enkephalin-induced analgesia (Kolesnikov et al., 1998).

Therefore, we have the *second* solution: If indeed enkephalinergic projections from VMH to PAG actually potentiate excitatory responses by NMDA-sensitive neurons that constitute output neurons from the PAG,

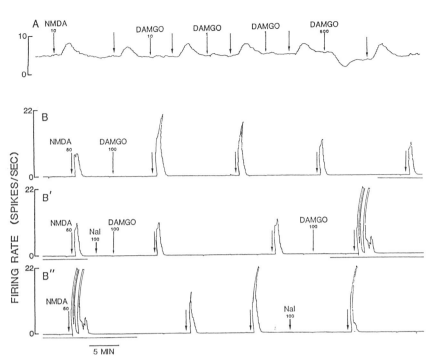

**Fig. 26.** Using single-unit recordings different effects of brief application of the MOR agonists [D-Ala, N-Me-Phe, Gly-ol]-enkephalin (DAMGO) can be observed in PAG neurons. In trace (*A*), NMDA (10 nm) and DAMGO (concentrations are indicated in nanomolar) are applied as a brief pulse through the recording chamber. NMDA applied at 10-min intervals (larger arrows) produces a consistent increase in firing rate. DAMGO applied at intervening time points has no effect on firing rate at low concentrations but promptly inhibits firing rate at 500 nM. This effect likely reflects a hyperpolarization produced by an opening of $K^+$ channels (Chieng and Christie, 1996). (*B, B′, B″*) Continuous recordings are shown from a single unit, with overlapping segments underlined. NMDA is applied at 10-min intervals (larger arrows). DAMGO, applied at an intervening time point, produces a robust increase in the response to NMDA that lasts tens of minutes. (*B′*) The opioid antagonist naloxone applied before DAMGO blocks the effect, and (*B″*), subsequently, reapplication of DAMGO again potentiates the response to applied NMDA.

then estrogen-induced enkephalin gene products could be properly understood as playing a facilitatory role in lordosis behavior circuitry.

Finally, the *third* solution—disinhibitory mechanisms between enkephalin neurons and GABA neurons—can be most easily understood from the electron microscopic immunocytochemistry presented in the next section.

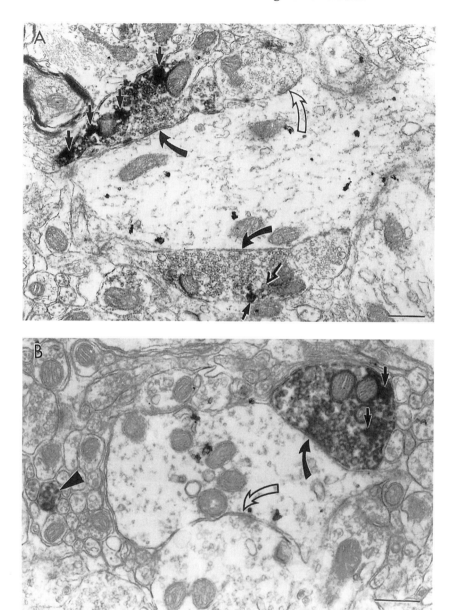

**Fig. 27.** In the ventromedial hypothalamic nucleus of the male rat, enkephalin-containing axons innervate GABAergic dendrites. (*A*) A dendrite containing immunogold particles for GABA receives a symmetric synapse from an enkephalin-labeled axon terminal (black curved arrows). An adjacent unlabeled axon terminal also forms a symmetric synapse on the GABA-labeled dendrite (open curved arrow). The upper-left enkephalin-labeled terminal contains several intensely labeled dense-

## D.  Intimate Relations among Inhibitory Systems

Particular relations among inhibitory systems in the forebrain represent the third solution of the aforementioned paradox. We have recently discovered a considerable number of special morphological relationships between two inhibitory systems: enkephalin and GABA (Figs. 27 and 28). Our recent work on hypothalamic mechanisms for reproductive behavioral systems, added to a solid literature from the hippocampus and from pain control circuitry in the PAG, shows us that *disinhibition* through the inhibition of GABA neurons by enkephalin neurons could represent an essential mechanism for estrogen-stimulated lordosis.

Enkephalin may function to dampen inhibition in the VMH. In addition to some notable cases whereby certain types of inhibitory neurons function to turn off other inhibitory neurons (Zieglgansberger et al., 1979), our data using ultrastructural immunocytochemistry illustrate how enkephalin could function to facilitate lordosis through disinhibition (Commons and Pfaff, 2000).

Opioids, such as enkephalin, characteristically inhibit neuronal activity, both by opening potassium channels and by decreasing intracellular accumulation of the signaling molecule, cyclic adenosine monophosphate (cAMP) (Duggan and North, 1983). However, there are several examples in which opioids produce neural excitation through a disinhibitory mechanism. The idea that opioids were involved in regulating inhibition was first established in studies of hippocampus (Zieglgansberger et al., 1979). The studies of these interneural relationships were facilitated by the laminar organization of the hippocampus. That is, the location and morphology, as well as electrophysiological characteristics, easily can be used to identify different types of neurons in the hippocampus (Amaral and Campbell, 1986). The principal excitatory cells, pyramidal cells, are lined up in a characteristic band and do not fire when recorded from in vitro. Inhibitory neurons, however, are distributed throughout the different areas of the hippocampus and are usually spon-

---

core vesicles (large straight arrow), whereas the lower terminal contains one labeled dense-core vesicle (large straight arrow) in addition to unlabeled dense-core vesicles (open arrow) Bar = 0.2 μm. (*B*) An enkephalin-labeled axon terminal contains peroxidase-labeling intense in two foci probably over dense-core vesicles (straight arrows). It forms a symmetric synapse (black curved arrow) on a GABA-labeled dendrite that also receives a synapse from an unlabeled axon terminal (open curved arrow). Also in the field is an enkephalin-labeled unmyelinated axon (arrowhead). Bar = 0.15 μm. (From Commons et al. 1999).

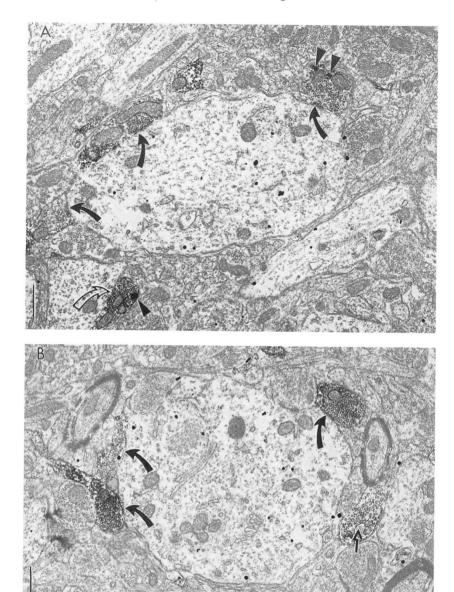

**Fig. 28.** Examples of GABAergic dendrites in the ventromedial nucleus of the female rat that receive multiple synaptic contacts from enkephalin-containing boutons. (*A*) Several enkephalin-immunoreactive axon terminals surrounding (curved black arrows) a dendrite containing immunogold labeling for GABA. Dense foci of labeling (arrowheads) likely correspond to enkephalin-containing dense-core vesicles. A neighboring enkephalin-immunoreactive axon terminal contacts (open ar-

taneously active. Interneurons are also unique in what types of neuropeptides they contain, and in the type of synaptic contacts they make.

When recording from hippocampal pyramidal cells, applied opioids produce bursts of firing. Investigators noted that the bursts were very similar to an excitation produced by drugs that blocked GABAergic synaptic activity (Zieglgansberger et al., 1979). Therefore, they proposed a model that GABAergic neurons in the hippocampus were inhibited by opioids, rather than directly exciting the pyramidal neurons. To support this idea, the investigators recorded from two neurons simultaneously: one spontaneously active, characteristic of a hippocampal interneuron, and one nonfiring, more typical of a hippocampal pyramidal cell. They infused an opioid over the cells, which both stopped the interneuron from firing and concomitantly produced a burst of firing in the pyramidal cell. To test whether the activation of the pyramidal cell was dependent on synaptic connections, they added $Mg^+$, which blocks synaptic transmission. The $Mg^+$ blocked the excitation of the pyramidal cell, strongly suggesting that it was mediated by disinhibition.

Subsequently, intracellular recordings demonstrated that opioids act through pre- and postsynaptic mechanisms to decrease the effectiveness of GABA neurons in the hippocampus (Cohen, Doze, and Madison, 1992). Acting at postsynaptic sites, opioids hyperpolarize the membrane potential of interneurons. Opioids also act to decrease the amount of GABA that interneurons can release when they fire action potentials, as well as to decrease the spontaneous release of the neurotransmitter. On the other hand, opioids do not affect the intrinsic membrane properties of pyramidal cells, including their sensitivity to GABA.

Anatomical observations have confirmed and extended these observations. For example, both $\mu$ and $\delta$ opioid receptors are present at postsynaptic sites, as well as in some axon terminals on hippocampal interneurons (Commons and Milner, 1997; Drake and Milner, 1999) (Fig. 29). Interestingly, separate populations of interneurons contain either $\mu$ or $\delta$ opioid receptors. Interneurons called basket cells (because their axons form a "basket" surrounding the pyramidal cell body) provide innervation surrounding the soma of pyramidal cells. They contain $\mu$ but not $\delta$ receptors (Drake and Milner, 1999). Interneurons that more specif-

---

row) an unlabeled dendritic profile. (*B*) A separate example of a GABA-labeled dendrite receiving several synaptic contacts from enkephalin-containing axon terminals (large black curved arrows); a nonsynaptic enkephalin axon is nearby (open arrow). From Commons and Pfaff, 2001.) Bars = 0.4 mm.

A

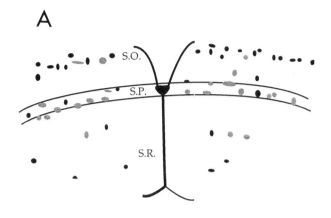

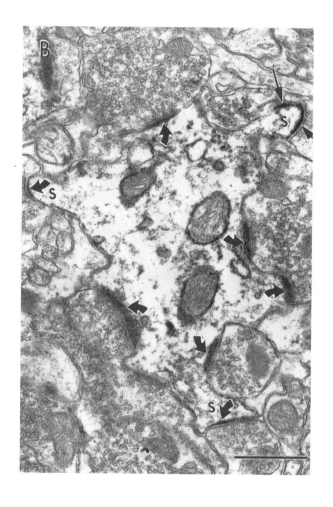

ically provide innervation farther out in the dendritic zone contain δ receptors but not μ receptors (Commons and Milner, 1997). This difference in distribution could explain why μ opioids can produce epileptic seizure in the hippocampus but δ opioids cannot (Siggins et al., 1986). Endogenous opioids act at both these sites to facilitate the activation of hippocampal pyramidal cells.

There are several areas besides the hippocampus where excitatory effects of opioids have been postulated to be mediated by disinhibition. Indeed, this appears as a repeated motif in several brain areas. One example of this would be in VTA, which is thought to be a critical area where opioids act to produce rewarding effects (Cameron, Wessondorf, and Williams, 1997). In the VTA, opioids produce an excitation of dopamine-containing cells that project to the forebrain. This excitation is thought to be a critical event in the pleasurable effects of opioids. Another example would be the PAG, where opioids activate descending neuronal pathways that function to inhibit the perception of pain, that is, antinociceptive pathways. The activation of these pathways is thought to underlie how opioids produce analgesia (Moreau and Fields, 1986; Depaulis, Morgan, and Liebeskind, 1987).

Could enkephalin function to modulate inhibitory interneurons in the VMH? Many of the axon terminals that come from estrogen-sensitive VMH neurons may make connections with local neurons within the VMH. In support of this idea, many of the synaptic connections in the VMH are from local neurons to local neurons (Nishizuka and Pfaff, 1989). Because of the strong expression of enkephalin in the VMH (Harlan et al., 1987), many of these axon collaterals likely contain enkephalin. In addition, other brain areas could send enkephalin-containing axon terminals to the VMH. For example, both the central nucleus of the amygdala and the PAG contain many enkephalin-producing cells and

---

**Fig. 29.** MOR and DOR immunoreactivity are present in interneuron populations in the hippocampal formation. (*A*) Location of interneurons containing immunoreactivity for MOR (gray dots) or DOR (black dots) is mapped with respect to different lamina of the CA1 region. Interneurons containing MOR cluster in the pyramidal cell layer (S.P.) while those containing DOR are most prevalent in stratum oriens (S.O.). A representative CA1 pyramidal cell and its dendritic arborizations is depicted. S.R., stratum radiatum. (*B*) In stratum oriens of CA1, DOR immunoreactivity is at postsynaptic sites in a spiny dendrite that is contacted by many unlabeled terminals (curved arrows) on both the shaft and spines (S). DOR labeling is diffuse but is most intense in one spine head at, and adjacent to, the postsynaptic region (straight arrows). Bars = 0.5 μm. (From Commons and Milner, 1997.)

project to the VMH (Harlan et al., 1987; Fahrbach, Morrell, and Pfaff, 1989).

Little is known about the function and neurocircuitry of GABA-containing neurons in the VMH. One reason for this is that immunohistochemical labeling of GABA neurons results in very diffuse staining throughout the VMH. Furthermore, it is quite difficult to identify different cell types in the VMH by either their location, morphology, or electrophysiological characteristics, as can be done in the hippocampus. Although poorly defined by light microscopy, ultrastructural analysis has revealed many GABA-containing neurons resident in the VMH (Commons et al., 1999).

The concentration of GABA in certain subsets of VMH neurons may reflect that these neurons use GABA as a neurotransmitter. It would make sense that many VMH neurons are GABAergic because there are many observations supporting the importance of this neurotransmitter in the function of VMH (McCarthy et al., 1991). Fiber and Etgen (1997) showed that GABA augments the release of norepinephrine. That fact could be relevant to the work of Kow et al. (1991) on the norepinephrine stimulation of reproductive behavior.

Indeed, GABAa receptor agonists facilitate lordosis when infused on the VMH (McCarthy, Malik, and Feder, 1990). There is a sexual dimorphism in GABA concentrations in the VMH, with males having more GABA than females in diestrus (Luine, Grattan, and Selmonoff, 1997). The work of Grattan and Selmanoff (1997) suggests lower turnover of GABA in the VMH in females compared with males; this could be due to inhibition of GABA neurons by enkephalin. In addition, estrogen increases GABA levels as well as GABA receptor levels in the mediobasal hypothalamus (Luine et al., 1999). There is also an increase in muscimol binding sites in the ventrolateral aspect of the VMH and the surrounding area with estrogen (O'Connor, Nock, and McEwen, 1988).

To detect whether enkephalin could function to modulate GABA in the VMH, we determined whether axon terminals containing enkephalin formed synapses with soma or dendrites containing GABA labeling. We labeled enkephalin and GABA using two different immunohistochemical-labeling methods that can be visualized using an electron microscope. With ultrastructural resolution, synaptic contacts also can be distinguished. These studies revealed that enkephalin-containing axon terminals provide a very strong innervation to GABA-labeled somata and dendrites within the VMH (Figs. 27 and 28). About one-third of enkephalin-containing terminals synapse on, or are adjacent to, a GABA-labeled neuronal profile (Commons et al., 1999, 2000). Furthermore, of the axon terminals that are seen surrounding GABA-labeled soma or dendrites, fully one-third contain enkephalin. Indeed, the frequency of these interactions far surpasses the number of the same types of synaptic connections in the hippocampus (Commons and Milner, 1996).

Therefore, anatomical evidence supports the notion that a role of enkephalin within the VMH may be to modulate the activity of GABA. One important role of GABAergic neurons may be to inhibit output from the VMH topically. Disinhibition of this excitatory output may facilitate lordosis. However, the role of GABA may be heterogeneous. For example, GABA also likely regulates a population of GABAergic VMH neurons. Understanding the interplay among these neurotransmitter systems will be critical for thoroughly understanding the behavioral output of this nucleus.

## E. Hypothalamic Projections

As introduced, two types of VMH projections also can help solve the paradox: (1) projections from the VMH to the POA, and (2) projections from the VMH to the PAG (Saper, Swanson, and Cowan, 1976; Krieger, Conrad, and Pfaff, 1979; Canteras, Simerly, and Swanson, 1994; Sato and Yamanouchi, 1999; Conrad and Pfaff, 1976a, 1976b). Since VMH enkephalin-producing neurons, crucial for lordosis (Pfaff and Sakuma 1979a, 1979b), project to the POA (Langub and Watson, 1992), they can inhibit, at that site, neural cell groups that are themselves inhibitory to lordosis behavior. In fact, ER-containing neurons, as well, project from the VMH to the POA (Turcotte and Blaustein, 1999).

There may be several neurochemicals, excitatory and inhibitory (Commons et al., 1999), within axons arising from the VMH, but enkephalin is abundant and well studied. Many VMH neurons express high levels of the enkephalin precursor gene, PPE (Harlan et al., 1987). PPE expression is remarkably correlated with the appearance of lordosis and is strongly stimulated by estrogen (Romano et al., 1988; Priest, Eckersell, and Micevych, 1995). Enkephalin is not only present in local axon terminals within the VMH, but also is transported to provide constituents of axon terminals that constitute the output of VMH in areas such as the MPOA and the PAG. This is evidenced by lesioning of the VMH, which creates a loss of enkephalin immunoreactivity in these areas (Hoffman et al., 1996).

## 1. To the MPOA

VMH neurons, including many estrogen-sensitive neurons, densely innervate the MPOA (Canteras, Simerly, and Swanson, 1994). With respect to lordosis, neuronal activity in the MPOA is negatively correlated with the appearance of the behavior (Pfaff, 1980). Indeed, the MPOA has the opposite effects on many aspects of autonomic and behavioral physiology compared to the medial or posterior hypothalamus. Activation of the MPOA can have a positive effect on locomotor activity, as in hopping and darting. The VMH, however, fosters immobility as during lordosis. It could be that a dynamic mutual inhibition is the basis of alternating increased locomotor activity during courtship and immobility for lordosis. The reciprocal connections between the VMH and MPOA provide an anatomical basis for this dynamic balance (Fahrbach, Morrell, and Pfaff., 1989). Furthermore, both the VMH and the MPOA can relay information through the PAG (Fahrbach, Morrell, and Pfaff, 1986; Canteras, Simerly, and Swanson, 1994), which is positioned to implement these alternative behaviors and accompanying autonomic changes (Mantyh, 1983a, 1983b).

Enkephalin within the axon terminals that project from the VMH to the MPOA is well positioned to mediate inhibition of the MPOA during mating behavior (Hoffman et al., 1996). Estrogen stimulates the release of endogenous opioids that bind $\mu$ opioid receptors (MORs) in the MPOA (Eckersell, Popper, and Micevych, 1998). This has been detected by observations on the subcellular distribution of MOR within neurons in the MPOA. That is, when MOR is bound to a ligand, MOR moves from the plasmalemmal surface of the neuron to the cell interior within endosomes (Sternini et al., 1996). This translocation can be visualized using confocal microscopy. Since the VMH likely sends an opioid-containing projection to the MPOA, and the MPOA has a negative influence on lordosis, activation of the opioid-containing pathway from the VMH to the MPOA could block the negative influence of the MPOA on the appearance of lordosis. Interestingly, opioids are reinforcing in the MPOA (Agmo and Gomez, 1993).

In view of these findings, it is of considerable interest that estrogen can increase the level of mRNA for $\mu$ opioid receptors in the VMH (Quinones-Jenab et al., 1997), and that progesterone can increase levels of $\mu$ opioid receptor mRNA in the POA as well as in the arcuate nucleus of estrogen-primed female rats (Petersen and LaFlamme, 1997). In addition, Eckersell, Popper, and Micevych (1998) used laser scanning microscopy to demonstrate that estrogen can induce the translocation of μ opioid receptor immunoreactivity from the membrane, in both the MPOA and the medial amygdala.

## 2. To the PAG

Enkephalinergic projections to the midbrain PAG (Yamano et al., 1986) also can help solve the paradox of estrogenic facilitation of enkephalinergic transcription in the VMH. Ultrastructural evidence suggests that ligands for μ opioid receptors (including enkephalin) and NMDA receptor ligands could act on common cell populations in the vlPAG (Fig. 30). In fact, in the manner of a neuromodulator, a μ receptor ligand can actually enhance responses to an NMDA receptor ligand in the midbrain PAG (Fig. 26) (Kow et al., 1998). Thus, it is surprising to conclude that enkephalinergic projections from the VMH to the vlPAG, acting through the affinity of enkephalin for μ opioid receptors, could actually enhance the output neurons from the PAG.

Projections to the midbrain PAG are important because of its known role in lordosis behavior (Pfaff, 1980; Behbehani, 1995) as well as its obvious role in the control of pain. Thus, it is significant that the same kinds of intimate relations among enkephalinergic neurons (reflected through μ receptors) and GABA neurons, which we reported earlier for hypothalamic systems, also occur in the midbrain PAG (Figs. 31 and 32). Indeed, this disinhibitory mechanism at a midbrain level of lordosis circuitry could provide the third solution to the paradox as easily as disinhibitory mechanisms in the hypothalamus, and the two levels of sexual behavior circuitry may, in fact, work in parallel. Local connections between midbrain PAG neurons have already been noted as being of potential physiological significance (Jansen et al., 1998). In concert with our electron microscopic immunocytochemical evidence (Figs. 31 and 32), Renno et al. (1999) noted colocalization of enkephalin and GABA in approximately 23 percent of axon terminals and synaptic terminations between GABA and enkephalin as well as between enkephalin and GABA. During recordings from rat hippocampal neurons, Lupica and colleagues (Lupica, 1995; Svoboda and Lupica, 1998), using electrophysiological techniques, discovered the ability of opioid peptide deposition to have a net excitatory effect on output through interneurons presumably using GABA.

The importance of the VMH projection to the PAG derives in part

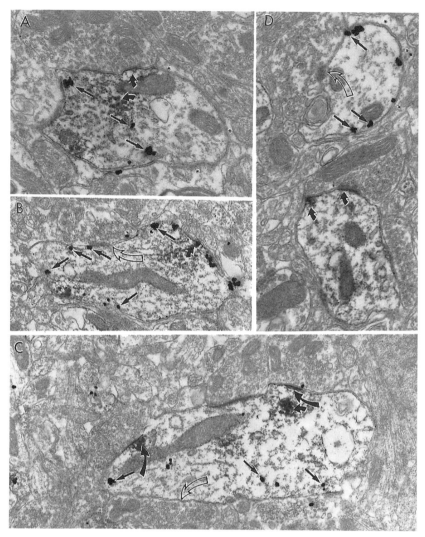

**Fig. 30.** In the PAG, neurons containing MOR (gold-silver labeling) are also those containing the NMDA receptor subunit, NR1 (peroxidase labeling). (*A*) A dendrite with both MOR labeling (straight arrows) and NR1 labeling (curved arrows) is shown. The dendrite is largely covered by an astrocytic process (asterisks). (*B*) Both MOR labeling (gold-silver, straight arrows) and NR1 labeling (black curved arrows) in a dendrite that receives one synaptic contact (open arrow) is shown. Note the robust plasmalemmal, but primarily nonsynaptic, localization of MOR labeling. (*C*) A dually labeled dendrite in which MOR labeling is present within the cytoplasm (straight arrows) and NR1 labeling is intense at two synapses (large black curved arrows), and in a patch of cytoplasm (small black curved arrow), is shown. Note that another synaptic contact is largely unlabeled (open arrow). (*D*) An example of den-

from its robustness (Saper, Swanson, and Cowan, 1976; Krieger, Conrad, and Pfaff, 1979; Canteras, Simerly, and Swanson, 1994) and in part from its functional roles. Initial interest in the PAG was driven by the observation that electrical stimulation of the area produced a powerful analgesia that was sufficient for surgery in a rat (Reynolds, 1969). PAG stimulation in humans can produce analgesia as well; however, patients also report unpleasant sensations such as intense foreboding or fear (Behbehani, 1995). Although the PAG has been most well studied for its role in gating painful sensations, it can effect many behaviors including the production of vocalizations, changes in heart rate, and motor behavior (running, freezing, immobility). Overall, the PAG appears to coordinate motor and autonomic responses in several types of survival situations, which are also often highly charged emotional events (Bandler and Keay, 1996).

To function in all these behaviors, the PAG receives topographically organized cognitive (cortical), limbic endocrine (hypothalamic), and sensory (spinal) information (Beitz, 1982c). Recently, an organizational scheme has been described in the PAG, that is, that the structure is composed of longitudinally organized columns (Bandler and Shipley, 1994). It has been argued that the lateral and ventrolateral areas of the PAG coordinate composite behaviors that represent either an active (escape or fight) response or passive responses (quiescence or immobility), respectively. The former responses would be generated in response to escapable threats, the latter to inescapable threats or in recovery from injury.

Much of our information on the PAG is associated with behaviors with a negative affective component, such as fearful and painful events. However, mating behavior mechanisms also relay through the PAG (Sakuma and Pfaff, 1979a, 1979b). The VMN gives its heaviest projection to the lateral PAG, but afferents from the VMN extend into the ventrolateral cell column as well (Canteras, Simerly, and Swanson, 1994). The lateral PAG descends to medullary reticular nuclei that directly innervate lumbosacral spinal areas that effect the muscle contractions arching the back—lordosis (Daniels, Miselis, and Flanagan-Cato, 1999). Both the lat-

---

drites single labeled for MOR or NR1 is shown. MOR labeling (straight arrows) is associated with the plasma membrane in areas directly opposed to an astrocytic process (asterisks) away from the synaptic junction (open arrow). In a nearby dendrite, NR1 labeling is most intense at and adjacent to a synaptic contact (curved black arrows). (From Commons Van Bockstaele, and Pfaff, 1999.)

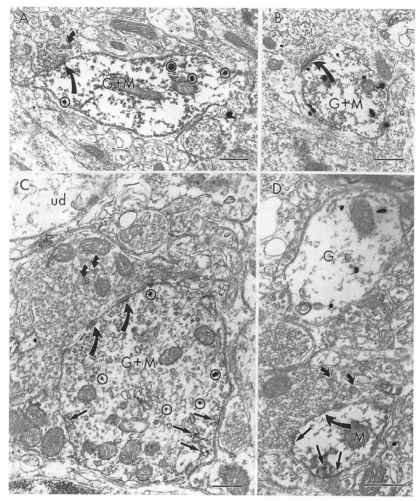

**Fig. 31.** GABAergic neurons (immunogold labeling) in the PAG contain MOR (per-oxidase labeling). (*A* and *B*) Two examples of small-caliber dendrites containing both GABA and MOR labeling (G+M) are shown. In these examples, MOR labeling is diffuse throughout the profile. These dendrites are contacted by single-axon terminals within the plane of section (large curved arrows). Dense-core vesicles are present in the presynaptic axon terminal in *A* (small curved arrow). (*C*) A large caliber dendrite containing both GABA and MOR labeling (G+M) is contacted (large curved arrows) by a large axon containing dense-core vesicles (small curved arrows) among synaptic vesicles. MOR labeling is diffuse but is most intense near or on the plasmamembrane (straight arrows). An unlabeled dendrite (ud) is nearby. (*D*) Two dendrites, each singly labeled for GABA (G) and MOR (M) are shown. MOR label-ing is largely associated with the plasmamembrane (straight arrows). Dense-core vesi-cles (small curved arrows) are present within the axon terminal presynaptic (large curved arrow) to this MOR-labeled dendrite. (From Commons et al., 2000.)

**Mu opioids may act at both pre- and postsynaptic sites to inhibit GABAergic neurotransmission in the PAG.**

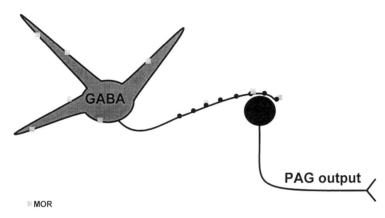

**Fig. 32.** Summary of potential sites whereby enkephalin could inhibit GABAergic interneurons in the PAG. The μ opioid receptor (MOR) is present in GABA-containing dendrites in the PAG, as well as on GABAergic axon terminals. Some of these terminals may directly contact PAG output neurons. (From Commons et al., 2000.)

eral and the ventrolateral PAG descend to the RVM, which has been implicated in modulating nociception through its projections to the spinal cord (Mantyh, 1983b) (Figs. 33 and 34).

While it may seem contradictory to portray the PAG as an area that both responds to stressful situations and functions in mating behavior, there are commonalities. First, the female has to decide whether the male rat is a threat to be escaped. This decision requires composite information arising from several modalities: cognitive, limbic and endocrine (hypothalamic), and sensory. Second, mating behavior constitutes not only courtship and lordosis but also requires cardiovascular and respiratory alterations associated with muscular effort. The PAG has the descending neurocircuitry capable of coordinating both these motor and autonomic responses (Beitz, 1982c; Mantyh, 1983a, 1983b).

How does the VMH projection to the PAG modulate activity in the PAG? The VMH projection in part contains enkephalin, and several aspects of how enkephalin acts to modulate neuronal activity in the PAG are known. Many opioids including enkephalin have net excitatory effects on PAG output; that is, the effects of opioids in the PAG mimic electrical stimulation or chemical stimulation with glutamate, or even GABA antagonists (which block the inhibitory control) (Behbehani, Jiang, and Chandler, 1990; Ogawa, Kow, and Pfaff, 1994). Like the hippocampus, and, as we have postulated, the VMH (Figs. 27 and 28), opioids are thought to act through disinhibition in the PAG (Figs. 31 and 32) (Moreau and Fields, 1986).

The effects of $\mu$ opioids can be mimicked by GABA antagonists and blocked by GABA agonists (Moreau and Fields, 1986; Depaulis, Morgan, and Liebeskind, 1987). Indeed, some neurons are under tonic inhibition in the PAG because GABA antagonists excite some cells (Ogawa, Kow, and Pfaff, 1994). Many anatomical features of the opioid-disinhibition circuit are found in the PAG. For example, enkephalin-containing terminals contact GABAergic neurons (Renno et al., 1999). A subpopulation of GABAergic neurons in the PAG contain MOR both pre-and post-synaptically (Commons et al., 1999). GABA neurons provide innervation to PAG output neurons (Williams and Beitz, 1990).

Although good evidence from many different approaches supports the idea that opioids inhibit inhibitory interneurons in the PAG, opioids probably act through other mechanisms in the PAG as well to facilitate the activation of descending neurons. Clues regarding the multiple potential mechanisms first arose from anatomical studies. Williams and Beitz (1990) observed that just as enkephalin-containing terminals innervated GABAergic neurons, they also provided a strong innervation of PAG output neurons. Furthermore, the degree of similarity between the GABAergic network in the PAG and the GABAergic network in the hippocampus was unclear. Approximately one-third of GABAergic terminals in the PAG synapsed on neurons that contained GABA labeling, a marked contrast with the situation in the hippocampus (Williams and Beitz, 1990; Commons et al., 2000).

Additional clues appeared while examining the distribution of MOR with respect to the NMDA receptor subunit, NR1, in the PAG (Commons et al., 1999). NMDA and morphine, applied directly to the PAG, have the same effect on analgesia (Jacquet, 1988). To explain how these two neurochemicals, which have opposite effects on neural activity but the same effect on PAG, output it was postulated that they act on largely separate populations of neurons (Behbehani and Fields, 1979). However, when examining where these receptors were localized in PAG neurons, there was a striking colocalization such that it seemed neurons that would be the most responsive to $\mu$ opioids would also be the most sensitive to NMDA (Commons et al., 1999). These observations were not strictly consistent with the disinhibition model.

While opioids have primarily inhibitory effects on neural activity through opening potassium channels, and also decrease cAMP levels (Duggan and North, 1983), there are examples in which opioids have excitatory effects as well (Chen and Huang, 1991). One example of this is that μ opioids are known to facilitate the activation of NMDA receptors by exogenous ligands. This effect is mediated through PKC. That is, it can be blocked by PKC blockers and mimicked by PKC activators (Chen and Huang, 1992). If opioids were to have excitatory effects on the same cells that NMDA activates, then that would explain how the two ligands have the same effect on analgesia.

Evidence to support the idea that μ opioids could facilitate excitation in PAG cells came from single-unit recordings using an in vitro slice preparation (Fig. 26) (Kow et al., 1998). Using this method, changes in firing rate can be recorded from PAG neurons as drugs are quickly applied in the bath. The application of enkephalin or DAMGO (the MOR agonist) produces several distinct effects in PAG neurons as determined with this method. First, in one population of neurons there is a prompt inhibition of firing rate. This effect likely reflects a hyperpolarization produced by an opening of $K^+$ channels. Second, in another population, brief application of DAMGO produces a prompt increase in firing rate. This effect is likely due to disinhibition. Third, we recently have observed another effect that lasts tens of minutes after DAMGO has been washed through the bath. That is, a brief application of DAMGO results in an increase in firing rate produced by applied NMDA. Since the prolonged modulatory effect of DAMGO on NMDA responses occurs in different cells and over a longer time course, it appears independent of the prompt inhibitory effect. Therefore, it appears analogous to the sustained modulator effect of DAMGO in the trigeminal nucleus caudalis, and not dependent on the inhibitory effect of DAMGO.

The idea that opioids could facilitate the activity of PAG neurons raises the possibility that enkephalin and NMDA together would facilitate the activity of neurons that descend to the medulla. To determine whether this would be plausible anatomically, MOR would have to be on neurons that descend to the medulla. And, indeed, a large portion of neurons that descend from the PAG contain MOR (Figs. 33 and 34) (Commons et al., 2000). Therefore, in the PAG, enkephalin originating in the VMH may act to disinhibit PAG neurons and also to facilitate their excitation directly (Fig. 35).

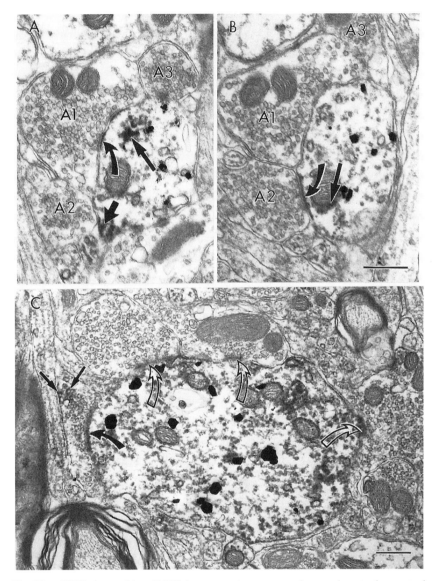

**Fig. 33.** MOR (peroxidase-DAB) is present in neurons that project to the ventral medial medulla identified with fluorogold (gold-silver) labeling. (*A* and *B*) Two sections of the same profile separated by approximately 300 nm are shown. Three axon terminals (A1, A2, and A3) surround a dually labeled dendrite. In *A*, axon A1 forms a synaptic contact (curved arrow) on the dendrite containing cytoplasmic MOR labeling (longer straight arrow) and MOR labeling associated with the plasmamembrane (short straight arrow). (*B*) The asymmetric synapse formed by axon terminal A2 is now visible (curved arrow) and MOR labeling is below that junction (straight

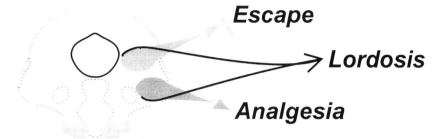

**Fig. 34.** Neural networks in the PAG control different behavioral outputs. Afferents to the lateral and ventrolateral PAG directly activate lordosis. Furthermore, these afferents also may act to (1) inhibit escape behavior and (2) dampen pain perception. Both of these effects would indirectly facilitate lordosis behavior.

## F. Gender Differences in Analgesia

There are impressive gender differences in opioid-controlled analgesia. Sex and pain are, indeed, intertwined. In the following sections, we review the literature, including a fair amount of work on sex differences in pain perception in humans. Most of the work elucidating the neuroanatomical, neurophysiological, neurochemical, pharmacological, molecular, and behavioral substrates of the functions of the endogenous opioid system in relation to analgesic processes has been carried out in rodents, and particularly young, postpubertal male rodents. It has become increasingly clear over the past decade that several organismic variables, particularly sex differences and aging differences, can account for specific changes in analgesic responsivity (see review in Bodnar, Romero, and Kramer [1988]). This section begins to outline the emerging picture that has come from this body of work in rodents but defers discussing these results with respect to human pain and analgesic states to a later section.

### 1. Gender Differences in Nociceptive Responses

Gender-specific effects in rats have been observed in reactivity to noxious stimuli with normal female rats displaying significantly lower shock-

---

arrow). (*C*) MOR and fluorogold labeling in a dendritic profile receiving several contacts from unlabeled axon terminals (open arrows) as well as one bouton containing light MOR immunoreactivity (straight black arrows point to presynaptic labeling, black curved arrow to the synapse) is shown. (From Commons et al., 2000.)

A

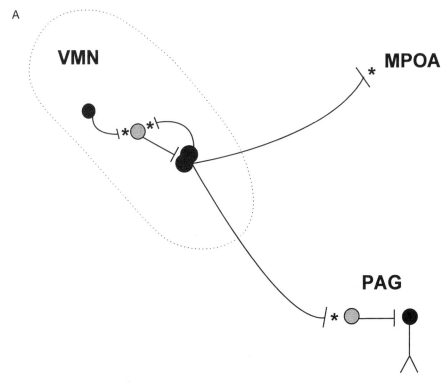

**Fig. 35.** (A) Solutions to the paradox for how enkephalin (*) facilitates lordosis through inhibiting neural activity are illustrated: (1) enkephalin may inhibit local inhibitory neurons in the VMN, facilitating output; (2) enkephalin may be present in projections from the VMN to the MPOA and block the inhibitory influence of MPOA on lordosis; and (3) enkephalin may project the PAG and act on local GABA-ergic neurons to disinhibit PAG output, thereby facilitating the lordosis response. Gray circles indicate local GABAergic neurons, and black circles indicate other types of neurons. (B) Enkephalin (*) containing afferents to the PAG may act two ways to facilitate PAG output: (1) enkephalin may inhibit GABAergic interneurons to disin-hibit descending projection neurons; and (2) enkephalin may act directly on de-scending neurons to facilitate the response to NMDA receptor ligands, directly fa-cilitating their excitation.

induced thresholds than males (Pare, 1969; Beatty and Beatty, 1970; Marks and Hobbs, 1972). Conversely, androgenized female rats display similar shock thresholds to normal male rats, and castrated male rats display similar shock thresholds to normal female rats (Beatty and Fessler, 1976, 1977). Changes in detection thresholds, but not noxious reactions, to shock are observed in female rats across the estrous cycle, with the great-est sensitivity occurring during periods of greatest estrogen activity

# Enkephalin may directly and indirectly modulate PAG output

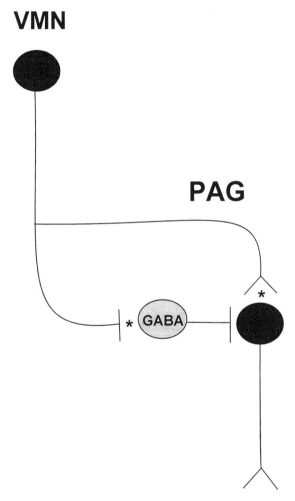

**Fig. 35.**  (*Continued*)

(Drury and Gold, 1978). By contrast, both intact and OVX female rats display slightly longer tail-flick latencies to radiant heat than intact and castrated male rats (Romero et al., 1987). As discussed later, the responsivity of rodents and humans to the detection and response to noxious stimuli is quite comparable as a function of sex differences, with females displaying greater reactivity than males across a series of dimensions. Sex differences appear more pronounced when evaluating analgesic processes, and the next section reviews sex differences in stress-induced analgesia.

## 2. Gender Differences in Stress-Induced Analgesia

Different parameters of stress-induced analgesia elicit opioid-mediated and nonopioid-mediated responses as indicated in opioid antagonist and morphine cross-tolerance studies. Thus, ICWS and CCWS analgesia have been characterized as, respectively, opioid and nonopioid by these and other criteria (see reviews in Bodnar [1984, 1986]). The magnitudes of both CCWS and ICWS analgesia were significantly lower in female rats as compared with either age- or weight-matched male rats, with reactivity to shock (jump thresholds) showing more consistency than reactivity to heat (tail-flick test) (Romero and Bodnar, 1986). Swim stress–induced analgesia was similar across the estrous cycle, which contrasted with greater estrous phase sensitivity for foot-shock-induced analgesia (Ryan and Maier, 1988). Gender differences in stress-induced analgesia have also been observed following exposure to either a predator or novelty as well as following immobilization stress: male mice displayed significantly greater levels of analgesia than female mice (Kavaliers and Innes, 1987a, 1988, 1992a; Kavaliers and Colwell, 1991).

The organizational and activational effects of gonadal hormones (e.g., Phoenix et al., 1959) were examined for gender differences in CCWS and ICWS analgesia by comparing sham-operated and castrated male rats with sham-operated and OVX female rats (Romero et al., 1987) (Fig. 36). Castrated male rats displayed significantly lower magnitudes of CCWS and ICWS analgesia than sham-operated male rats but the levels were similar to those of sham-operated female rats. OVX female rats displayed significantly smaller magnitudes of CCWS and ICWS analgesia than sham-operated female rats. Similar patterns of gender and gonadectomy differences were observed following swim analgesia in mice (Wong, 1987) and inescapable tail shock in rats (Ryan and Maier, 1988). These gender and gonadectomy differences in analgesia were unrelated to any swim-induced hypothermic or activity changes, suggesting that gonadal steroids facilitated these stress-related responses. Steroid replacement therapy can determine whether this treatment ameliorates the deficit in gonadectomized animals and further enhances the stress-induced analgesic effects in intact animals (Fig. 36). Steroid replacement therapy with testosterone reversed the deficits in CCWS and ICWS analgesia observed in castrated male rats and OVX female rats (Romero et al., 1988); however, it produced minimal effects in intact male and female rats, resulting in the ability of both intact and castrated male rats treated with testosterone to display significantly greater magnitudes of CCWS and ICWS analgesia than both intact and OVX female rats. Steroid replacement therapy with estradiol benzoate reinstated swim-induced analgesia in castrated mice (Wong, 1988), and estrogen replacement reinstated tail-shock-induced analgesia in OVX rats (Ryan and Maier, 1988). The effectiveness of gonadal steroid replacement in adult gonadectomized but not intact rats to reinstate swim stress–induced analgesia suggests that the activational effects of gonadal steroids play a necessary role for the elaboration of the analgesic response in each gender, and that their full expression is necessary for the observation of sex differences in the magnitudes of both opioid- and nonopioid-mediated forms of swim stress–induced analgesia.

Kavaliers and coworkers have taken a wide-ranging ethological approach in demonstrating sex differences in stress-induced analgesia. Male deer mice displayed significantly greater immobilization- and novelty-induced analgesia than females in mainland and island populations (Kavaliers and Innes, 1987a, 1988). The sex differences in immobilization-induced analgesia also affected their responsivity to NMDA antagonism or to the neuropeptides NPFF and Tyr-MIF-1: analgesia in male deer mice was blocked, whereas the lesser analgesia in female deer mice was only blunted (Lipa and Kavaliers, 1990; Kavaliers and Innes, 1992a, 1992b). Whereas prolonged exposure to a cat produced a naloxone-sensitive and 5-HT-insensitive analgesia that was greater in male mice, brief exposure to a cat produced a naloxone-insensitive and 5-HT-sensitive analgesia that was greater in female mice (Kavaliers and Colwell, 1991). A similar pattern of temporal and pharmacological effects was observed in analgesia elicited in male and female meadow voles following prolonged or brief exposure to their predator, a garter snake (Saksida, Galen, and Kavaliers, 1993). Further, brief exposure to biting flies elicited naloxone-sensitive analgesia that was significantly greater in male deer mice, whereas prolonged exposure to biting flies produced a nonopioid and NMDA-sen-

A

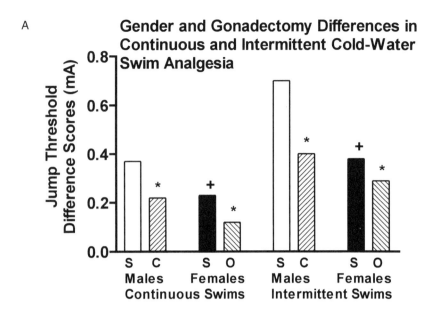

B

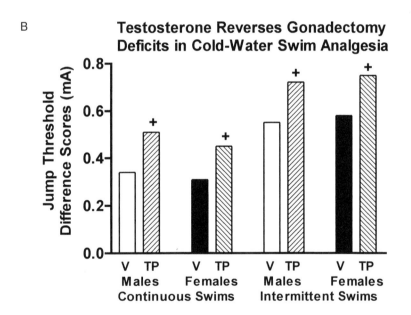

**Fig. 36.** (*A*) Adult male rats displayed significantly greater magnitudes of analgesia following both nonopioid-mediated CCWSs and opioid-mediated ICWSs than adult female rats regardless of whether the male rats were matched for age or body weight with the females. Adult gonadectomy, performed 1 mo before testing to as-

sitive analgesia approximately equivilent in both sexes (Kavaliers, Colwell, and Choleras, 1998). These data indicate the range and ethological scope of these sex differences in analgesic responses elicited by environmental stimuli (for a review see Choleris and Kavaliers [1999]).

Sex differences in stress-induced analgesia are not limited just to the magnitude and duration of the effects, but also to their underlying pharmacology. Thus, the respective opioid- and nonopioid-mediated analgesic responses to ICWS and CCWS initially determined in male rats was altered in females such that naloxone significantly reduced ICWS analgesia in sham-operated and castrated males, but not in age-matched, sham-operated, or OVX females (Romero, Kepler, and Bodnar, 1988b). CCWS analgesia was unaffected by naloxone pretreatment in intact and gonadectomized males and female rats. Thus, an animal's gender determined the opioid character of ICWS analgesia and may be related to the existence of a separate gender-specific and estrogen-dependent analgesic system in female mice in which a nonopioid ($15°C$) form of swim-induced analgesia in mice is both sexually dimorphic and hormonally dependent (Mogil et al., 1993). Maximal levels of swim-stress analgesia were observed in sham-operated and gonadectomized male and female rats, which was significantly reduced by NMDA antagonism in intact and castrated male rats, but not in intact female rats. OVX females displayed the "male-like" pattern of NMDA-mediated swim-stress analgesia, but estrogen replacement in these OVX animals restored NMDA-insensitive swim-stress analgesia indicative of intact female animals. Neonatal androgenization of female mice with testosterone produced male-like NMDA-mediated swim stress analgesia during adulthood even in the presence of estrogen (Sternberg et al., 1995). Moreover, female mice engaged in breeding displayed an opiate-insensitive and MK-801-insensitive swim-induced analgesia that was lesser in magnitude than in male mice. Tests of deer mice during photoperiodically induced breeding (cycling behav-

---

sess the activational effects of gonadal hormones, significantly reduced both CCWS and ICWS analgesia in castrated male rats relative to sham-operated controls, and in OVX female rats relative to sham-operated controls. (From Romero and Bodnar, 1986; Romero et al., 1987.) (*B*) The diminished analgesic response to CCWSs and ICWSs observed in castrated male rats and OVX female rats was reinstated by pretreatment with testosterone propionate (TP) for 14 days relative to sesame oil vehicle (V) injections. S, Sham-operated controls; C, castrated males; O, OVX females. *, Significant reductions in analgesic responses of animals receiving adult gonadectomy relative to animals receiving sham surgeries; +, significant gender differences in analgesic responses. (From Romero et al., 1988.)

iors) and nonbreeding (anestrous) conditions showed that "nonbreeding" females displayed greater NMDA-sensitive swim analgesia than "breeding" females (Kavaliers and Galea, 1995). This suggests the existence of an estrogen-dependent pain-inhibitory system in female mice in which estrogen allows analgesic expression through an unknown pathway that involves neither opioid nor NMDA receptors and must occur in the absence of testosterone during early organizational ontogeny.

Another instance in which the sex of the animal changes the underlying pharmacology of analgesic processes occurred in a hormonally stimulated pregnancy model (Liu and Gintzler, 2000). Castrated male rats, like OVX females, display analgesia following combined and prolonged treatment with 17β-estradiol and progesterone, but not with either treatment alone. In female rats, this analgesic response depends on a spinal interaction between δ and κ but not μ receptors (Dawson-Basoa and Gintzler, 1996, 1997, 1998). By contrast, hormonally stimulated analgesia in castrated male rats depends on a spinal interaction between κ and μ but not δ receptors (Liu and Gintzler, 2000). Furthermore, hormonally stimulated analgesia is reduced by intrathecal $\alpha_2$-adrenoceptor antagonists in OVX females (Liu and Gintzler, 1999), but not castrated males (Liu and Gintzler, 2000).

Finally, mouse strain differences also alter gender differences in stress-induced analgesia. Swim-stress analgesia elicited in C57BL/6J and DBA/2J females was insensitive to either naloxone or MK-801 (Mogil and Belknap, 1997). However, C57BL/6J males displayed swim-stress analgesia that was blocked by MK-801, but not naltrexone, while this analgesia in DBA/2J males was blocked by naltrexone, but not MK-801. A sex-specific quantitative trait locus on chromosome 8 mediating nonopioid stress-induced analgesia was identified in female mice that accounts for one-fourth of the overall trait variance in this sex (Mogil et al., 1997b). Habituation-induced analgesia induced by repeated saline injections and hot-plate tests was reduced by $\delta_2$ but not by μ or κ antagonists in a strain- and gender-dependent manner: DBA/2J males > C57BL/6J males > DBA/2J females > C57BL/6J females (Mogil et al., 1997a).

### 3. Aging Differences in Stress-Induced Analgesia

Aging has been identified as another factor in determining the magnitude and potency of different forms of stress-induced analgesia. Although basal nociceptive thresholds decrease as a function of age in rodents (Gordon, Scobie, and Frankl, 1978; Chan and Lai, 1982), other studies report increases in basal nociceptive thresholds across different ages (Pare, 1969; Hess, Joseph, and Roth, 1981). The analgesic responses to glucoprivation and CCWS have been analyzed for aging effects. Our laboratory (Kramer, Sperber, and Bodnar, 1985) found significant age-dependent reductions in the magnitude of glucoprivic analgesia (Fig. 37). Parallel reductions in the magnitude of glucoprivic hyperphagia across age groups suggested orderly alterations in glucoprivic responses rather than a specific alteration in pain-inhibitory systems. Whereas four younger cohorts (4–19 mo of age) of female rats displayed similar patterns of CCWS analgesia on a thermal nociceptive measure, the oldest (24 mo) group of females showed a decrement in CCWS analgesia (Kramer and Bodnar, 1986b). A gradual decline in the magnitude of CCWS analgesia on a shock test was observed in the three older groups of female rats. Age-dependent CCWS hypothermia was dissociated from CCWS analgesia since the three older groups of female rats showed potentiated CCWS hypothermia normally indicative of greater analgesic responsivity. Such age-related declines in stress-induced analgesia were also observed for both opioid and nonopioid forms of foot-shock analgesia (Hamm and Knisely, 1985), but not for analgesia induced by food deprivation (Hamm and Knisely, 1985, 1986; Hamm, Knisely, and Watson, 1986). Moreover, swim-stress analgesia increased as a function of age in male rats (Hamm, Knisely, and Waton, 1986), indicating an important analgesic interaction between aging and gender.

## 4. Gender Differences in Morphine Analgesia

Both normal female and castrated male rats and mice display significantly less magnitudes of morphine analgesia than normal male rats and mice on somatic as well as visceral measures (Chatterjee et al., 1982; Badillo-Martinez et al., 1984a; Kavaliers and Innes, 1987b, 1987c; Baamonde, Hidalgo, and Andres-Trelles, 1989; Candido et al., 1992). As observed for restraint stress-induced analgesia, the magnitude of morphine analgesia displayed sex-dependent reductions by pretreatment with either NMDA antagonists or the peptide agonists, NPFF and Tyr-MIF-1 (Lipa and Kavaliers, 1990; Kavaliers and Innes, 1992a, 1992b). These sex differences in morphine analgesia in deer mice were evident from weaning to adulthood, suggesting that organizational, rather than activational, effects of gonadal hormones were responsible (Kavaliers and Innes, 1990). Chronic naloxone treatment produces analgesia through upregulation of opioid receptors that is greater in male than in female

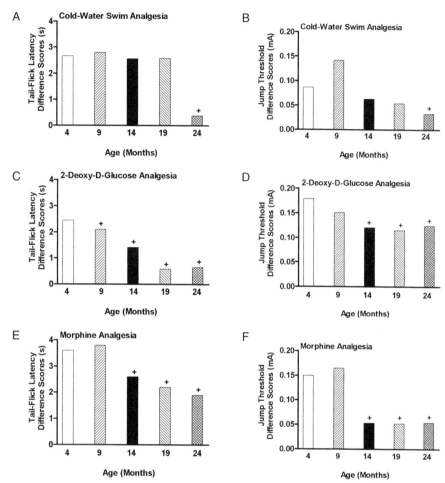

**Fig. 37.** Adult female rats displayed significant age-dependent changes in analgesia as measured by the tail-flick (*A, C,* and *E*) and jump (*B, D,* and *F*) tests following CCWSs (*A* and *B*), 2-deoxy-D-glucose glucoprivation (*C* and *D*), and morphine (*E* and *F*). For the nonopioid-mediated analgesic response to CCWS analgesia, the magnitude of the response was maintained across age groups up to 19 mo of age and then declined in 24-mo-old rats (Kramer and Bodnar, 1986a). By contrast, analgesia displayed more orderly age-related decrements across groups following administration of either the opioid-mediated stressor, 2-deoxy-D-glucose, or morphine (Kramer, Sperber, and Bodnar, 1985; Kramer and Bodnar, 1986a). +, Significant age-dependent alterations in different analgesic responses relative to 4-mo-old rats.

deer mice (Kavaliers and Innes, 1993a). The greatest sensitivity to morphine analgesia across the estrous cycle appears to occur during the late diestrous phase (Banerjee, Chatterjee, and Ghosh, 1983). However, μ-selective sex differences are not universally observed, especially following systemic fentanyl and buprenorphine (Bartok and Craft, 1997).

Our laboratory (Kepler et al., 1989) presumed that this gender-specific action on morphine analgesia would be centrally mediated and found that sham-operated male rats displayed significantly greater and more potent ventricular morphine analgesia than sham-operated females (Fig. 38). Whereas intact male rats displayed near-maximal analgesia at a 5-μg morphine dose, intact females displayed only moderate analgesia at doses as high as 40 μg. In contrast to the effects of gonadectomy on stress-induced analgesia, castrated males and OVX females displayed only minor alterations in the magnitude and potency of morphine analgesia relative to intact controls. Further, whereas the proestrous phase significantly increased the magnitude of ventricular morphine analgesia on the tail-flick test, the jump test failed to display significant differences across the estrous cycle.

In assessing strain differences, the AKR/J, C57BL/6J, and SWR/J strains displayed ventricular morphine analgesia that was significantly greater in males than in females (Kest, Wilson, and Mogil, 1999). By contrast, female CBA/J mice were fivefold more sensitive to supraspinal morphine analgesia than males. The suggestion that the presence and direction of sex differences in morphine analgesia varied as a function of strain was tested in three outbred rat and three outbred mouse strains (Mogil et al., 2000). Systemic morphine analgesia was greater in male Long-Evans and Sprague-Dawley rats relative to corresponding females, but morphine analgesia was equivalent in both sexes in Wistar Kyoto rats. Further, systemic morphine analgesia was greater in male CD-1 and ND-4 mice relative to corresponding females, but morphine analgesia was equivilent in both sexes in Swiss Webster mice. An overall role of the relationship between genotype and pain and analgesia recently has been assessed (Mogil and Basbaum, 2000).

Sex and aging variables interact in systemic morphine analgesia. As indicated earlier, female and male rats display respective age-related decrements and enhancements in swim-stress analgesia (Hamm, Knisely, and Watson, 1986; Kramer and Bodnar, 1986b). Age-related decreases in opiate receptors (Messing et al., 1980; Hess, Joseph, and Roth, 1981) and β-endorphin, leu-enkephalin, and met-enkephalin levels (Gambert et al., 1980; Barden et al., 1981; Dupont et al., 1981; Gambert, 1981) have been

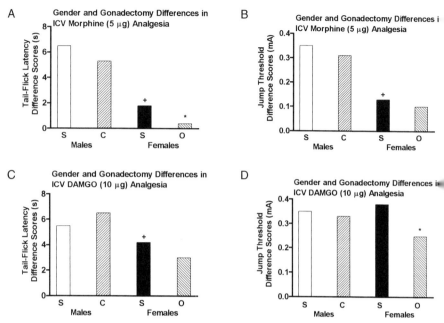

**Fig. 38.** (*A* and *B*) Adult male rats displayed significantly greater magnitudes of analgesia following intracerebroventricular (ICV) administration of morphine than adult female rats. This effect was observed across a range of morphine doses and occurred regardless of the estrous phase of the female rat. In contrast to different forms of cold-water swim analgesia, adult gonadectomy failed to alter morphine analgesia significantly following ventricular administration (Kepler et al., 1989). (*C* and *D*) Adult male rats displayed significantly greater magnitudes of analgesia only on the tail-flick test following intracerebroventricular administration of the μ-selective opioid agonist, DAMGO, than adult female rats. Again, adult gonadectomy failed to alter markedly DAMGO analgesia following ventricular administration (Kepler et al., 1991). S, Sham-operated controls; C, castrated males; O, OVX females. *, Significant reductions in analgesic responses of animals receiving adult gonadectomy relative to animals receiving sham surgeries; +, significant gender differences in analgesic responses.

noted. Early studies of aging effects on morphine analgesia alternatively have reported reductions (Webster, Shuster, and Eleftheriou, 1976; Chan and Lai, 1982; Kavaliers, Hirst, and Teskey, 1983), increases (Saunders et al., 1974), or both reductions and increases (Spratto and Dorio, 1978) because of different definitions of aged rodents, pain tests employed, and the limited use of dose-and time-response functions. Age-related decrements in morphine analgesia in female rats were observed (Kramer and Bodnar, 1986a) that could not be attributed to alterations in baseline nociceptive thresholds or hyperthermia. These age-related effects on opioid analgesia appeared to be selective to morphine since β-endorphin analgesia was intact across female rats ranging from 8 to 30 mo of age (Romero and Bodnar, 1987). Examination of the relationship among sex differences, gonadectomy, and age differences on morphine analgesia revealed significant decreases in the $ED_{50}$ for morphine analgesia in intact sham-operated males, and significant increases in the $ED_{50}$ for morphine analgesia in intact sham-operated females (Islam, Cooper, and Bodnar, 1993). Importantly, adult ovariectomy at 3 mo of age eliminates this age-dependent decline in female rats. These latter data may link the estrogen-dependent pain-inhibitory system in female rats proposed for stress-induced analgesia (Mogil et al., 1993; Sternberg et al., 1995) to these age-dependent effects.

In a study of Cicero, Nock, and Meyer (1996), the generalizability of sex differences in morphine analgesia was extended to several pain tests such that male rats displayed greater systemic morphine analgesia than females on the tail-flick, hot-plate, and abdominal constriction tests that could not be attributed to differences in serum levels of morphine. A further study by these researchers (Cicero et al., 1997) found that sex differences in systemic morphine or alfentanil analgesia could not be explained by morphine pharmacokinetics because there were similar linear relationships in males and females between the morphine dose and blood and brain morphine levels at that interval (60 min) when the sex differences were most apparent, suggesting the importance of inherent sex differences in the sensitivity of the brain of males and females to morphine.

## 5. Gender Differences in Opioid Analgesia

While sex differences appear to be somewhat consistent for morphine analgesia in animal studies following central or peripheral administration, available data are less clear regarding systemic or ventricular administration of selective opioid receptor subtype agonists. Ventricular administration of the selective μ opioid agonist, DAMGO, produced significantly greater analgesia on the tail-flick test in intact male than in intact female rats, but evaluation of the analgesic potency of DAMGO among sham-operated and gonadectomized male and female rats did not yield significant gender differences (Kepler et al., 1991) (Fig. 38). Again, adult gonadectomy failed to alter either the pattern or magnitude of DAMGO analgesia. Moreover, analgesia induced by the δ receptor agonist, DSLET, failed to differ among intact and gonadectomized male and female rats.

Consistent sex differences in κ opioid receptor analgesia have been observed with U50488H or the enkephalinase inhibitor, SCH34826, increasing latencies more in male than in female deer mice throughout development (Kavaliers and Innes, 1987b, 1990, 1993b). Indeed, these κ-mediated sex-dependent effects have been used to explain gender differences in "nonopioid" stress-induced analgesia. Mice exposed to the odor of a weasel predator displayed an analgesia that was significantly greater in male mice and was insensitive to general or κ opioid antagonists. NMDA antagonism blocked this response as well as analgesia elicited by the κ agonist, U69593, in male but not female mice, suggesting that nonopioid (actually non-μ) forms of stress-induced analgesia in male but not female mice may be mediated through κ opioid and NMDA mechanisms (Kavaliers and Choleris, 1997). However, the ability of all opioid receptor subtypes to produce sex-dependent analgesia has been called into question.

Bartok and Craft (1997) found that ventricular $\delta_1$ (DPDPE) or $\delta_2$ (Delt II) agonists produced significantly greater hot-plate latencies in male than in female rats with no sex differences in agonist potencies. Indeed, U69593 produced greater peak analgesia in females on the hot-plate test with no sex differences noted thereafter. Analgesia elicited by nonopioid agents displays dichotomous gender difference effects. Whereas analgesia elicited by either the $\alpha_2$-adrenergic agonist, clonidine, the muscarinic cholinergic agonist, pilocarpine, or cocaine is greater in male than female rats, the magnitude of nicotine analgesia was similar in males and females (Kiefel and Bodnar, 1992; Craft and Milholland, 1998). Although the ventricular approach has clarified sex differences in morphine analgesia in animals quite independent of pharmacokinetic factors, ventricular injection techniques are subject to the drawback of agonist diffusion from those relevant sites in which sex differences in analgesia may occur. By contrast, ventricular morphine with a greater half-life than opioid peptides and peptide analogues may reach these relevant sites despite diffusion and thereby activate sex-dependent effects.

## 6. The vlPAG as a Potential Site Mediating Gender Differences in Opioid Analgesia

The vlPAG is a strong candidate to mediate gender differences in opioid analgesia because this site (1) elicits supraspinal opioid analgesia through μ and, secondarily, $\delta_2$ opioid receptor subtypes (e.g., Fang, Fields, and Lee, 1986; Jensen and Yaksh, 1986a, 1986b; Bodnar et al., 1988; Smith et al., 1988; Rossi, Pasternak, and Bodnar, 1993; Rossi et al., 1991); (2) mediates sex- and hormone-dependent reproductive behaviors, particularly through its interaction with estradiol-concentrating hypothalamic loci; and (3) mediates interactions between sex hormones and opioid peptides, particularly control of transcription of the PPE gene by estradiol. While evidence supporting the first point has been thoroughly reviewed, research relevant to the second and third points follows.

Sexual and reproductive behaviors in the female rat are mediated in part in the PAG. The neural circuitry controlling lordosis in the female rat has identified the mesencephalic PAG as an important relay station in the mediation of this response (see review in Pfaff and Schwartz-Giblin [1988]). The PAG receives descending afferents from the VMH and MPOA (e.g., Krieger, Conrad, and Pfaff, 1979), where many estrogen-concentrating cells are located (e.g., Pfaff and Keiner, 1973; Fahrbach, Morrell, and Pfaff, 1986). Those compartments of the PAG sensitive to gonadal steroids in turn project to the medial medullary reticular formation and the spinal cord, including sites that respond to pressure stimuli adequate for eliciting lordosis (e.g., Kow and Pfaff, 1982). These sites have considerable overlap with medullary analgesic loci in the RVM. Electrical stimulation of the PAG facilitates lordosis (e.g., Sakuma and Pfaff, 1979a), whereas lesions placed in the PAG suppress this response (Sakuma and Pfaff, 1979b; Riskind and Moss, 1983). Further functional evidence indicates that whereas the PAG is necessary for lordosis, the VMH provides tonic hormone-related facilitation of this response by such means as estrogen-induced peptides that are transported to the PAG to make it ready for the induction of lordosis (e.g., Mobbs et al., 1988).

The role of opioids in reproductive responses and their reliance on the PAG is derived from several lines of evidence. First, ventricular administration of specific μ (DAMGO), δ (DPDPE), and κ (U50488H) opioid agonists alters lordosis differentially (Pfaus and Pfaff, 1992). DAMGO significantly reduced lordosis in female OVX rats treated with estradiol benzoate and progesterone, but not in rats primed with estradiol benzoate alone. By contrast, DPDPE and U50488H each facilitated lordosis in both treatment groups. Thus, facilitation of lordosis by δ and κ agonists appears independent of progesterone treatment, whereas the inhibitory effects of μ agonists on lordosis require the presence of progesterone. Second, the opioid peptide, met-enkephalin, appears to be one of the estrogen-induced polypeptides that mediate the effects of estrogen on lordosis. The ventrolateral subdivision of the VMH is the site where (1) estrogen most effectively affects lordosis (Davis, McEwen, and Pfaff, 1979), (2) estrogen increases PPE mRNA levels (Romano et al., 1988, 1990), and (3) estrogen is colocalized with met-enkephalin (Akesson and Micevych, 1991). Induction of PPE gene expression by estrogen treatment occurs within 1 h (Romano et al., 1989). Met-enkephalin typically produces inhibitory actions on PAG neurons in electrophysiological studies using a tissue slice preparation (Ogawa, Kow, and Pfaff, 1994). Enkephalin's inhibitory action on PAG neurons was dose dependently increased in estrogen-primed females, but not OVX females. Indeed, several studies have provided definitive links between estrogen modulation of proenkephalin gene products and the role that the endogenous opioids play in sex- and hormone-specific reproductive behaviors (e.g., Lauber and Pfaff, 1990; Lauber et al., 1990; Pfaff et al., 1996). Therefore, these data indicate that the PAG is an important link in gender-specific reproductive behavior, that the VMH and MPOA possess important links with the PAG by which estrogen can modulate reproductive behavior, that opioid receptor subtypes differentially mediate reproductive behavior, and that a compelling link among anatomical, neuroendocrinological, and neurochemical substrates may be estrogen's increase in PPE mRNA gene expression and its subsequent effects on PAG neurons. These hypotheses have begun to be systematically tested, and the next section summarizes recent results.

## 7. The vlPAG and RVM Mediate Gender Differences in Opioid Analgesia

Both sham-operated ($ED_{50}$ = 1.20–1.60 µg) and castrated ($ED_{50}$ = 1.08–1.09 µg) male rats displayed similar magnitudes and potencies of vlPAG morphine-induced analgesia that were significantly greater than sham-operated female rats tested during the estrous phase ($ED_{50}$ > 50 µg) or OVX female rats ($ED_{50}$ = 1.98–2.51 µg) (Krzanowska and Bodnar, 1999) (Fig. 39), a pattern similar to that for ventricular morphine or DAMGO (Kepler et al., 1989, 1991). Similarly, RVM morphine analgesia was significantly greater in male rats than in female rats (Boyer, Morgan, and Kraft, 1998), indicating that the vlPAG and RVM, two critical supraspinal opioid analgesia sites (for reviews see Fields and Basbaum [1978]; Basbaum and Fields [1984]), display sex differences following morphine treatment. Morphine's analgesic magnitude in male rats approached

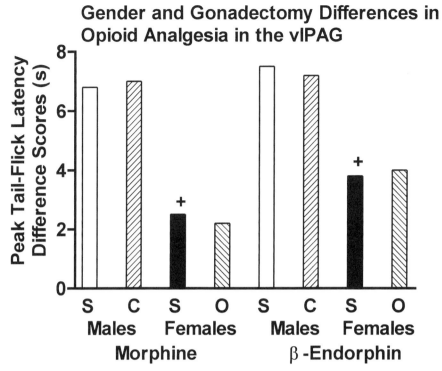

**Fig. 39.** The vlPAG was confirmed as a primary site where gender differences in opioid analgesia could be observed. Adult male rats displayed significantly greater magnitudes of analgesia following intracerebral administration of either morphine or β-endorphin than adult female rats. This effect was observed across a range of opioid agonist doses, and, again, adult gonadectomy failed to alter significantly morphine or β-endorphin analgesia following vlPAG administration. The persistent gender differences observed for both agonists are quite striking given the dissociable mechanisms of action observed for morphine and β-endorphin, including within the vlPAG. S, Sham-operated controls; C, castrated males; O, ovs females. +, Significant gender differences in analgesic responses. (From Krzanowska and Bodnar, 1999, 2000a.)

cutoff values, yet only moderate levels were observed in females even at higher morphine doses. vlPAG morphine analgesia, like ventricular actions, was largely unaffected by male gonadectomy, indicating that circulating and activational male sex steroid hormones are not crucial for the full expression of this response. By contrast, OVX females displayed greater vlPAG morphine analgesia than intact females, particularly at higher doses resulting in a leftward shift in morphine's potency in OVX relative to intact, estrous-tested females. One potential mechanism is

that individual or combined sex steroid hormones (estradiol and progesterone) might act to inhibit vlPAG morphine analgesia in intact females. The PAG mediates interactions between sex hormones and opioid peptides, particularly control of transcription of the PPE gene by
estradiol (Lauber and Pfaff, 1990; Lauber et al., 1990; Pfaff et al., 1996),
with the latter increasing hypothalamic PPE mRNA at a rapid rate in females relative to males as a function of estradiol level across the estrous
cycle (Romano et al., 1988, 1989, 1990). Although estradiol-induced increases in PPE mRNA in females might logically suggest greater availability of endogenous opioids, and therefore greater analgesic magnitudes, such sex steroid–induced modulatory effects may produce changes
in responsivity to exogenously applied opiates, particularly at high doses,
in that stage in which such steroids are maximally active (estrous) and
in a site (the PAG) that would be maximally sensitive to such changes.

Progesterone itself produces analgesia in OVX female rats (McCarthy,
Malik, and Feder, 1990; Frye, Bock, and Kanarek, 1992) and male mice
(Kavaliers and Wiebe, 1987), although only at moderate, but not high,
levels (Frye et al., 1996). Chronic progesterone decreases analgesia
elicited by sucrose exposure (Frye, Bock, and Kanarek, 1992). Moreover,
OVX female rats displayed a pattern similar to pregnancy-induced analgesia only if combined but not individual progesterone and estrogen
treatment mimicked blood profile levels corresponding to late pregnancy and parturition (Dawson-Basoa and Gintzler, 1993). By contrast,
estrogen-primed OVX rats displayed an enhanced analgesic response to
vaginal stimulation that was blocked by concurrent administration of progesterone (Rothfield, Gross, and Watkins, 1985). Furthermore, concurrent administration of progesterone dampened analgesia of the testosterone metabolite, 3β-androstanediol (Frye et al., 1996), and blocked the
estrogen-induced facilitation of intrathecal muscimol analgesia (McCarthy,
Malik, and Feder, 1990). Therefore, in pregnancy, progesterone and estrogen synergize estrogen to facilitate analgesia, whereas in vaginal stimulation, progesterone inhibits estrogen-mediated analgesia. Since female rats in the estrous phase display muted vlPAG morphine analgesia,
this suggests that high circulating levels of both gonadal hormones may
produce mutually antagonistic effects (Krzanowska and Bodnar, 1999).

To test the generalizability of vlPAG gender differences, analgesia induced by other opioid agonists, β-endorphin and D-Pro$^2$-Endomorphin-
2, were examined (Krzanowska and Bodnar, 2000a, 2000b; Krzanowska
et al., 2000b). vlPAG β-endorphin produced gender-dependent analgesia with both sham-operated and castrated male rats displaying greater

potency (twofold) than sham-operated, estrous-phase, or OVX female rats (Fig. 39). A surprising opposite pattern was observed with vlPAG D-Pro$^2$-Endomorphin-2 analgesia in that estrous-phase, sham-operated, and OVX female rats displayed significantly more potent analgesia than sham-operated and castrated male rats on the tail-flick but not jump test. Interestingly, male rats appeared to be behaviorally more active following vlPAG D-Pro$^2$-Endomorphin-2, relative to either vlPAG morphine or β-endorphin. In formally assessing this activity, alterations in ambulatory or stereotypic activity induced by β-endorphin were sensitive to gender differences (Krzanowska and Bodnar, 2000b), with commonly shared initial reductions following vlPAG β-endorphin but subsequent increases in ambulatory activity observed in males, but not females. By contrast, activity changes following vlPAG D-Pro$^2$-Endomorphin-2 were not gender dependent (Krzanowska et al., 2000b). However, males and females differed in behavioral activation following vlPAG D-Pro$^2$-Endomorphin-2 such that significantly greater grooming, seizure, barrel-roll, and explosive running behaviors occurred in males, but not females, during those intervals when males displayed "hyperalgesic" responses on the tail-flick test. Thus, any attributed sex differences following vlPAG D-Pro$^2$-Endomorphin-2 do not appear to affect nociceptive processing per se but, rather, appear to be an epiphenomenon of behavioral activation.

Another study (Tershner, Mitchell, and Fields, 2000) found that female rats displayed significantly greater vlPAG DAMGO-induced analgesia than male rats on the tail-flick test. A methodological difference from previous studies was that the females in this study were not assessed for the phase of the estrous cycle, whereas the females in the morphine and β-endorphin studies were only tested during the estrous phase of the cycle. Since there were differences in the magnitude of intracerebroventricular morphine analgesia as a function of the estrous phase (Kepler et al., 1989), this factor deserves further consideration.

Another additional point deserves note. Whereas male rats displayed reduced vlPAG morphine analgesia following RVM pretreatment with a κ agonist (Pan, Tershner, and Fields, 1997), female rats displayed analgesia following RVM administration of the κ agonist U69593 (Tershner, Mitchell, and Fields, 2000). This raises the question: Would RVM κ agonists produce sexually dimorphic responses on on-cell and off-cell activity?

A recent series of studies (Krzanowska et al., 2000a) investigated whether organizational effects (Phoenix et al., 1959) of circulating sex hormones were responsible for sex differences observed for vlPAG morphine analgesia. The activational effects of female gonadal hormones suppress vlPAG morphine analgesia since adult OVX females showed more potent vlPAG morphine analgesia than sham-operated controls, whereas adult castrated males failed to differ from sham-operated controls in vlPAG morphine analgesia. To assess organizational gonadal hormone effects, 1-day-old male rat pups received sham or castration surgery, and 1-day-old female rat pups received either vehicle or testosterone proprionate (200 μg/kg). Sham males displayed significantly greater vlPAG morphine analgesia in adulthood than sham females (Fig. 40). Neonatally androgenized female rats displayed a marked increase in the po-

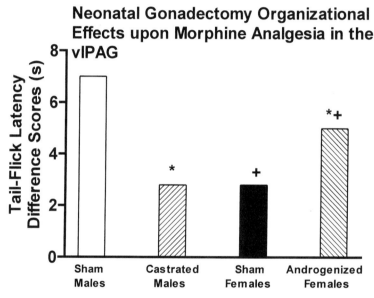

**Fig. 40.** The gender differences in morphine analgesia in the vlPAG appeared not to be sensitive to adult gonadectomy, suggesting a minor role for the activational effects of gonadal hormones. By contrast, gender differences in morphine analgesia in the vlPAG appeared very sensitive to the organizational effects of gonadal hormones as studied using sham-operated and gonadectomized animals at 1 day of age, and subsequently tested in adulthood. Neonatally castrated rats displayed significantly smaller magnitudes of morphine analgesia in the vlPAG than sham-operated males, and roughly equivalent effects to those observed in sham-injected females. Neonatal female rats, androgenized with testosterone propionate, displayed significantly greater magnitudes of morphine analgesia in the vlPAG than sham-injected females, and their analgesic responses approximated those observed in sham-operated males. *, Significant alterations in analgesic responses of animals receiving adult gonadectomy relative to animals receiving sham surgeries; +, significant gender differences in analgesic responses. (From Krzanowska et al., 2000b.)

tency, magnitude, and duration of vlPAG morphine analgesia relative to vehicle-treated females that was approximately equivalent to that observed for sham-operated males. Correspondingly, neonatally castrated male rats displayed a marked decrease in the potency, magnitude, and duration of vlPAG morphine analgesia relative to sham-operated males that was approximately equivalent to that observed for vehicle-treated females. Thus, perinatal organizational influences of gonadal hormones are integral for the successful formation of sex differences in opioid agonist–mediated analgesia.

## 8. Human Studies Investigating Sex Differences in Pain and Analgesia

The previous sections have detailed the considerable evidence demonstrating sex differences in either nociceptive or analgesic responding in experimental animals. This section briefly reviews controversial issues relating human sex differences in pain and analgesic processes, which have been the source of debate as well as interest in the past five years. Berkley's (1997) comprehensive review, which we follow, examining those situations under which women display lower thresholds for pain or rate painful stimuli as either more intense or less tolerable. Several critical variables have been identified, including incidence assessment of endogenous pain states between males and females, differences in somatic nociceptive stimuli in experimental studies, the relationship between sex organs and sex differences in human pain, and the relationship between sex hormone fluctuations and sex differences in human pain. Pain-related disorders with a higher incidence in males than females include migraine, cluster and posttraumatic headaches, duodenal ulcers, abdominal migraines, postherpetic neuralgia, hemophiliac anthropathy, and ankylosing spondylitis, whereas females display a higher incidence for chronic tension, postdural puncture and cervicogenic types of headaches, tic douloureux, temporomandibular joint disorder, occipital neuralgia, Raynaud disease, carpal tunnel syndrome, and sympathetic dystrophy (Merskey and Bogduk, 1994; Berkley, 1997). By contrast, no sex prevelance has been established for acute tension headache, cluster-tic syndrome, secondary trigeminal neuralgia, vagus nerve neuralgia, chronic gastric ulcers, Crohn disease, and carcinoma or divertiular diseases of the colon. Thus, there is a failure to identify clearly defined sex-specific syndromes displaying a clear prevalence of pain-related disorders. Rather, sex differences appear attributable to nonsensory factors, including greater health awareness in females relative to males, types of assessments, and temporal and social factors (Bush et al., 1993; Marcus, 1995; Unruh, 1996; Berkley, 1997). Although coping behaviors have been hypothesized to play an important role in human sex differences to painful stimuli, strong gender differences were not found in the use of coping strategies in the appraisal of pain (Unruh, Ritchie, and Merskey, 1999).

Human sex differences in experimentally induced pain states have typically used psychophysical procedures and indicate that female subjects display lower thresholds to noxious stimuli, rate similar noxious stimuli as more painful, and display less tolerance for intense noxious stimuli (Feine et al., 1991; Lautenbacher and Rollman, 1993; Maixner and Humphrey, 1993; Ellermeier and Westphal, 1994; Fillingim and Maixner, 1995; Berkley, 1997). However, meta-analysis determined that many smaller studies reporting diminished sex differences had inadequate sample sizes independent of whether the investigators were examining threshold measures of pain or tolerance measures of pain (Riley et al., 1998). Moreover, more robust effects occurred when the type of experimentally induced pain involved pressure or electrical stimulation, and weaker effects were elicited when thermal measures of pain were employed. Human sex differences became more apparent when multiple measures of responding were employed, including determination of thermal pain threshold and tolerance, thermal discrimination real-time estimates of heat pulses, and temporal summation of thermal pain (Fillingim et al., 1998). Female subjects displayed lower thermal pain thresholds and tolerance and greater temporal summation of thermal pain than male subjects; the magnitude of this thermal effect was consistent in females across the menstrual cycle (Fillingim et al., 1997). Summated rather than punctate thermal stimuli are better in assessing sex differences, since female human subjects displayed lower thermal pain thresholds when a method-of-levels procedure was employed independent of whether the rise time of the thermal stimulus was fast or slow (Fillingim, Edwards, and Powell, 1999; Fillingim, Maddux, and Shackelford, 1999).

A relationship exists between the sensitivity to experimentally induced pain and chronic clinically relevant pain disorders including tuberoinfundibular disorders (Maixner et al., 1995, 1998), fibromyalgia (Lautenbacher and Rollman, 1993; Mountz et al., 1995), and headache (Langemark et al., 1989, 1993), such that patients with these disorders will display lower thresholds to experimentally induced pain through either altered physiological or psychological sequelae. Some of the sex differences observed in female relative to male subjects in experimentally induced pain might be attributed to a greater incidence of clinical pain-related symptoms experienced by the subjects 1 mo before study. Female subjects displayed lower thermal thresholds and experienced more clinical pain symptoms than males. Whereas that subgroup of female subjects displaying greater than average numbers of clinical pain symptoms showed greater sensitivity to thermal stimuli than females with fewer clinical pain symptoms, the relationship between the amount of clinical pain reports and experimental thermal thresholds failed to differ in males. This suggests that the experimental pain responses may be more clinically relevant in females relative to males (Fillingim, Edwards, and Powell, 1999). This relationship was also found for familial pain histories in which females with greater familial pain histories displayed greater sensitivity to experimentally induced thermal stimuli than females with fewer familial pain histories (Fillingim, Edwards, and Powell, 2000). By contrast, the experimentally induced thermal thresholds of male subjects, which were higher than those of females, failed to vary as a function of familial pain history. Thus, even in experimental pain situations, factors outside of the testing situation can affect the magnitude of the sex differences in human pain thresholds. Nonetheless, it still appears that the experimental animal literature, the human clinical pain literature, and human experimental pain literature indicate lower pain thresholds and lower pain tolerance in females.

The congruence between animal and human models of pain does not appear to extend to opioid analgesia studies. The most pronounced sex differences in animal studies were for morphine analgesia following systemic, ventricular, and intracerebral administration into the vlPAG and RVM with male rodents displaying significantly greater analgesic responses than female (Chatterjee et al., 1982; Badillo-Martinez et al., 1984a; Baamonde, Hidalgo, and Andres-Trelles, 1989; Kepler et al., 1989; Candido et al., 1992; Islam, Cooper and Bodnar, 1993; Cicero, Nock, and Meyer, 1996, 1997; Boyer, Morgan, and Craft, 1998; Krzanowska and Bodnar, 1999). Conversely, although analgesia induced by other opioid receptor subtypes, especially the κ subtype, produced significant sex differences in mice and deer mice with males showing greater effects (Kavaliers and Innes, 1990; Kavaliers and Choleris, 1997), little or no significant effects were observed in rats (Bartok and Craft, 1997). Interestingly, human studies using a model of postoperative dental pain demonstrate that the κ receptor elicits significantly greater magnitudes of analgesia in female subjects. Thus, κ agonists (pentazocine, nalbuphine, butorphanol) produced significantly greater levels of analgesia in female relative to male subjects following removal of the third molar teeth in a dose-dependent manner (Gear et al., 1996a, 1996b, 1999). By contrast, higher (20 mg) and lower (5 mg) nalbuphine doses, respectively, increased and decreased postoperative pain thresholds in male subjects, suggesting a potential form of sex-dependent antianalgesic mechanism.

These findings that human females display greater analgesia than males following κ agonists may be related to the role for dynorphin and κ opioid receptors in the mediation of pregnancy-induced analgesia in rats (Gintzler, 1980; Sander, Portoghes, and Gintzler, 1988; Medina, Dawson-Basoa, and Gintzler, 1993; Dawson-Basoa and Gintzler, 1996). Indeed, these spinally mediated responses may suggest a potential site of action for the human since κ opioid receptors in the dorsal horn vary as a function of the estrous cycle with the greatest densities of receptors observed during proestrous and estrous (Chang, Archer, and Drake, 2000). A great deal of further research is necessary to rectify sex differences in analgesia in animal and human studies.

## G. Pain and Sex: Similarity of Reproductive Behavior and Analgesia Ascending and Descending Pathways

Large portions of the lordosis behavior circuit, stretching all the way from the lumbar spinal cord, through the brain stem, midbrain PAG, and VMH, are very similar to both ascending pain-inducing and descending centrifugal analgesia circuitry. Estrogen-binding neurons in the vlPAG represent one example of this generalization, which is supported in detail in the following section.

### 1. Relation of Strong Somatosensory Inputs to Sexual Behavior, Pain, and Aggression

Small female rodents typically used for laboratory neurobiology are in nature prey. Somatosensory contacts from a variety of other animal species would signal immediate danger. Therefore, the predominant behavior in a female not hormonally prepared for reproduction would be to avoid such somatosensory input even as the female would avoid pain. Indeed, the primary behavioral reactions of sexually unreceptive female rats to somatosensory contacts, unless handled and soothed frequently, feature vigorous escape, immediate aggression, and anguished vocalizations. Likewise, sexually unreceptive female hamsters show fierce aggression toward the approaching male, attacking in a manner that eventually could lead to the death of the male even following light cutaneous input (Floody and Pfaff, 1974, 1977a, 1977b; Floody, Pfaff, and Lewis, 1977). Female rodents will learn and display arbitrarily chosen operant responses to gain access to a male (e.g., Matthews et al., 1997). In turn, the stronger the somatosensory input by the male toward the female in such encounters, the longer it will take the female to allow the male access once again (Bermant, 1961; Peirce and Nuttall, 1961). Pain signaling is an essentially ascending circuitry, whereas sexual behavior and analgesic responses are essentially a descending circuitry.

Physiological experiments indicate negative correlations between the intensity of responses to pain and sex (Table 8). Following bilateral VMH lesions, lordosis behavior failed even when estrogen and progesterone

**Table 8. Opposite Effects of Manipulations on Lordosis Behavior and Responses to Pain**

| Experiment | Effect on Lordosis Behavior | Effect on Response to Pain | Reference |
|---|---|---|---|
| Cutaneous pinch | ↓ | ↑ | Komisaruk et al., 1976; Komisaruk, 1974 |
| Vaginal probe | ↑ | ↓ | |
| Ventromedial hypothalamic lesions | | | |
|   Female rats | ↓ | ↑ (irritability) | Mathews and Edwards, 1977; Hetherington and Ranson, 1943 |
|   Female hamsters | ↓ | ↑ (aggression) | Malsbury, Kow, and Pfaff, 1977 |
| Midbrain central gray | | | |
|   Lesions | ↓ | ↑ | Sakuma and Pfaff, 1979b |
|   Stimulation | ↑ | ↓ | Sakuma and Pfaff, 1979a; Mayer and Liebeskind, 1974; Geisler and Liebeskind, 1976 |
| Septal lesions | Normal, then ↑ | ↑, then normal | Nance et al., 1974, 1975; McGinnis et al., 1978; Nance et al., 1977 |

*Note:* References are in Pfaff [1982].

were supplied in rodents (Mathews and Edwards, 1977; Sakuma and Pfaff, 1979b) and hamsters (Malsbury, Kou, and Pfaff, 1977). By contrast, instead of showing sexual behavior, such females react to cutaneous inputs of either aggressive males or experimenters with aversive or "irritable" responses, avoiding contact or attacking violently (Malsbury, Kow, and Pfaff, 1977; Mathews and Edwards, 1977). Thus, females with VMH lesions behave as though male the somatosensory input is aversive.

The inverse correlation of responsivity to pain and the female's response to sexually relevant stimuli receives support from physiological experiments in the midbrain PAG. Such neurons receive a strong input from the VMH (Conrad and Pfaff, 1976a, 1976b; Krieger, Conrad, and Pfaff, 1979). Female rats with bilateral lesions of either the midbrain PAG or of hypothalamic inputs to the PAG fail to show lordosis. In such animals, when cutaneous stimulation is applied by an experimenter in an attempt to elicit lordosis, females may avoid contact vigorously as though the cutaneous input were aversive (Kow, Conrad and Pfaff, unpublished observations). Thus, when testing for reproductive behavior, the predominant response shown by the female to such somatosensory input is typically a response to noxious stimulation.

Interestingly, such hypothalamic-PAG circuitry is also relevant to the full expression of feline aggression. Beginning with the classic studies of John Flynn (Egger and Flynn, 1962, 1963; Flynn et al., 1970; Bandler and Flynn, 1971, 1972; Flynn, Edwards, and Bandler, 1971; Wasman and Flynn, 1962), a lateral hypothalamic circuit was identified, that, when stimulated, would elicit stereotyped responses related to predatory attack. By contrast, and of particular relevance to the present discussion, VMH stimulation elicited powerful, affective aggressive responses, particularly against intraspecies conspecifics in cats. This relationship emerges as important if the female rat will activate sexual responses in the VMH when the female is sexually receptive, and aggressive responses in the VMH when the female is not. A further intriguing interactive response is, of course, the ability to elicit analgesic responses following stimulation of the VMH (Balagura and Ralph, 1973; Rose, 1974). Obviously, analgesic responses would be of heuristic value under conditions of lordotic behavior to maintain the full integrity of the sexual response, as well as aggressive behavior to maintain the possibility of continued aggressive responses or to allow withdrawal responses independent of injury. The relationship between aggressive encounters and the elicitation of analgesic responses has been summarized such that analgesia occurs following aggressive encounters, especially in the submissive, losing animal (Miczek, Thompson, and Schuster, 1982). Such a situation might be mimicked by a sexually unreceptive female rat that comes in contact with a sexually aggressive male rat. Although this aggression-elicited analgesic response is opioid mediated, but independent of pituitary mechanisms (Thompson et al., 1988), it is probably dependent on the VMH for its full expression, since μ opioid receptors and β-endorphin are decreased in the VMN following aggressive encounters in submissive animals (Miczek, Thompson, and Schuster, 1986; Kulling et al., 1988).

As indicated previously, one of our laboratories established that an intrinsic connection between the VMH and the PAG is essential for the maintenance of lordotic behavior. Continuing with the possibility that the VMH might act as an "oscillator" between lordotic behavior during sexual receptivity in hormone-replete females and aggressive behavior during various types of nonreceptivity, it would follow that a similar pathway should exist for aggressive behavior. Siegel and colleagues have identified such an aggression-inducing pathway in cats that uses VMH-PAG connections such that the PAG, like the VMH, supports affective forms of aggression (Wasman and Flynn, 1962; Bandler, 1977; Shaikh et al., 1985). Hence, stimulation of the medial midbrain tegmentum, including the PAG, facilitated affective attack and concurrently suppressed quiet, predatory attack (Shaikh et al., 1985). Stimulation of dorsal PAG placements suppressed hypothalamically elicited flight behavior, whereas stimulation of ventral PAG placements facilitated hypothalamically elicited flight behavior (Brutus, Shaikh, and Siegel, 1985). Interestingly, naloxone blocked the inhibition by PAG placements on hypothalamically elicited hissing behavior but failed to alter the facilitation by other PAG placements on hypothalamically elicited hissing behavior (Pott, Kramer, and Siegel, 1987), coinciding with naloxone's ability to facilitate affective defense behavior elicited from the feline VMH (Brutus and Siegel, 1989). Furthermore, the ability of selective μ and δ opioid agonists to suppress affective defense elicited by stimulation of the cat PAG was blocked by selective μ and δ opioid antagonists, respectively, indicating a localized inhibition of attack behavior by opioids within the PAG (Shaikh, Lu, and Siegel, 1991a, 1991b). Moreover, microinjection of 5-HT-1A and 5-HT-2/1C receptor agonists respectively suppresses and facilitates aggressive responses elicited from the PAG (Shaikh, DeLanerolle, and Siegel, 1997).

As a side point, the aggressive behavior elicited by PAG stimulation is inhibited by activity in the medial amygdala through μ but not δ opioid synapses (Shaikh, Lu, and Siegel, 1991b). By contrast, medial amygdaloid stimulation facilitates aggression elicited by PAG stimulation through substance P and NMDA synapses (Shaikh, Steinberg, and Siegel, 1993; Shaikh, Schubert, and Siegel, 1994; Schubert, Shaikh, and Siegel, 1996). The amygdaloid modulation of aggression as elicited from the PAG involves interactions between the VMH and the lateral hypothalamus such that stimulation of the medial amygdala activates VMH substance P receptors, which triggers an inhibitory GABAergic mechanism from the medial to the lateral hypothalamus (Han, Shaikh, and Siegel, 1996a; Han, Brutus, adn Siegel, 1996). This effect suppresses predatory behaviors and allows for the full expression of affective aggressive responses.

In conclusion, affective aggressive behavior elicited from the VMH of cats appears to extend to the dorsal aspects of the PAG, which is mediated by NMDA receptor activation. In turn, aggression induced by either the VMH or the PAG is initiated by the medial and central amygdala through substance P and GABAergic and opioid synapses (Siegel, Schubert, and Shaikh, 1997). As indicated throughout the book, such a system demonstrates remarkable similarities with systems mediating both lordotic and analgesic responses.

To establish the third component of the triad of lordotic-aggressive-analgesic responses as essential aspects of VMH-PAG mechanisms, our laboratories are collaborating to examine whether VMH or other hypothalamic lesions alter the integrity of opioid analgesia elicited from the vlPAG, especially the pattern of analgesic sex differences.

As reviewed (Pfaff, 1982), a number of experimental manipulations that reduce the noxious aspects of otherwise painful stimulation actually can improve lordosis behavior performance, and vice versa (Table 8). The most dramatic experiment was that in which vaginal pressure reduced responses to pain but increased the effectiveness of subsequent stimuli in producing lordosis behavior (Komisaruk 1974). Although spinally mediated mechanisms are important in this response, a supraspinal component has been identified using spinally transected animals (Watkins et al., 1984b). However, the role of the VMH-PAG axis in this form of antinociception is not yet well characterized.

Thus, we have proposed that one aspect of estrogen-influenced action on hypothalamic neurons allows these neurons to cause the animal to treat somatosensory inputs as sexually relevant, when otherwise they would have been treated as noxious. That is, neurons in the VMH and PAG, at the top of the lordosis circuit, would militate against competing responses such as limb flexions and bilaterally asymmetric trunk movements (see Pfaff and Lewis [1974]) and other avoidance responses that prevent the lordosis reflex from appearing. The actions of estrogens followed by progestins, working through the hypothalamus and its projections to the midbrain PAG, reduce the irritability and pain avoidance responses typical of unhandled OVX animals. The overlap of ascending and descending neuroanatomical pathways between lordosis behavior circuitry and analgesia-producing pathways supports the interpretation of a hormone-dependent gating of strong somatosensory inputs to allow reproductive behaviors.

## 2. Neuroanatomical Overlap of Lordosis Circuitry and SPA Pathways

### a. Ascending Pathways

Once lordosis-relevant cutaneous inputs reach the spinal cord, they are processed in the dorsal horns before being sent to higher levels of the CNS. We note that estrogen-binding neurons are found in Rexed layer II and that stimulus requirements for certain neurons in Rexed layer V fit those for lordosis behavior as a whole. Matching this, painful inputs from high threshold receptors are processed through wide dynamic range neurons (Mendell, 1966; Wall, 1967), and Rexed layers II and V are particularly important for pain signaling (Kolmodin and Skogland, 1960; Christensen and Perl, 1970; Willis, 1988).

Lordosis behavior–relevant ascending input travels through the anterolateral columns to the lower brain stem (Pfaff, 1980). Similarly, physiologically identified wide dynamic range types of cells innervate a wide range of ascending spinothalmic, spinoreticular, and spinomesencephalic fiber tracts through which nociceptive processing ascends through the spinal cord (Willis, 1985). These tracts include the spinothalamic tract, which has been studied for thermal and mechanical stimuli at its source in dorsal horn interneurons through its termination in the ventral posterior nucleus of the thalamus as well as the lateral cervical nucleus and the nucleus gracilis (Willis, 1986). As Willis (1988) succinctly indicates, these cells respond to different classes of stimuli: (1) intermediate-to-high noxious mechanical and thermal stimuli corresponding to warning, (2) high-intensity stimuli corresponding to damage, and (3) rapidly adapting mechanoreceptors. In addition to the spinothalamic tract, other nociceptive-processing pathways include the spinocervical tract, the postsynaptic dorsal column pathway, and the lamina I component of the spinomesencephalic tract, which access the medullary reticular formation, particularly the NRM and NRGC, as well as the vlPAG (Willis, 1985, 1988; Hylden, Hayashi, and Bennett, 1986; Duggan and Morton, 1988). Thus, painful inputs from pressure, temperature, pressure, inflammation, or arthritis travel through the lateral columns in the phylogenetically ancient paleospinothalamic as well as other tracts. These comprise upward-going axons carrying nociceptive information and encompass at least six subsystems, roughly equivalent to the classic "Flexor Reflex Afferent" pathways described initially by Lundberg, Malmgren, and Schomburg (1987), who studied mechanoreceptor responses interacting with primary afferent inputs to produce either facilitation or inhibition. Thus, both ascending lordosis behavior and ascending spinal pain pathways have an important spinoreticular component.

Both lordosis mechanisms and pain signaling include important inputs to neurons of the medullary reticular formation that are specifically sensitive to cutaneous stimuli of the sort that trigger lordosis (Kow and Pfaff, 1982). In parallel, neurons in the NRM and in the NRGC of the RVM respond specifically to painful inputs (see reviews in Willis [1985, 1988]; Duggan and Morton [1988]). Another similarity between the lordosis and pain systems is that both types of functions require transections on both sides for complete elimination, indicating that some sensory pathways are crossed in both cases, and some are not.

More parallels are observed in the midbrain. Even as pathways important for lordosis reach the lateral columns of the midbrain PAG (Pfaff, 1980), pain-bearing ascending pathways trigger responses in the midbrain PAG, especially its lateral, ventrolateral, and ventral portions (for reviews see Willis [1985, 1988]; Duggan and Morton [1988]). In addition to activating relevant descending centrifugal pain-inhibitory mechanisms, these pathways activate the LC, SN, and VTA, which are each sensitive to neurophysiologically confirmed nociceptive stimuli (Cedarbaum and Aghajanian, 1978; Gysling and Wang, 1983; Rasmussen and Jacobs, 1986; Rasmussen, Morilak, and Jacobs, 1986; Abercrombie and Jacobs, 1987a, 1987b; Morilak, Fornal, and Jacobs, 1987; Jacobs et al., 1991; Horvitz, 1997). Classically described dimensions of pain divided into sensory-discriminative, motivational-affective, and cognitive-attentional aspects apply, in that the LC is intimately involved in attentional mechanisms, especially the aspect of sustained attention responsible for vigilance and orienting (see reviews in Aston-Jones et al. [1986, 1996]). By contrast, the SN and VTA are involved in motivational-affective responses (e.g., Wise, 1982; Wise and Bozarth, 1987; Wise and Rompre, 1989). In turn, those cortical and subcortical mechanisms activated by the LC mediate many forms of conditioning, including, potentially, analgesia elicited by classical conditioning methods (autoanalgesia; e.g., Chance, 1980), which, in turn, are dependent on the vlPAG and the NRM for full expression. Similarly, among the motivational-affective mechanisms stimulated by noxious activation of the VTA may be the amygdala. As indicated earlier, opioid analgesia elicited from the amygdala is, in turn, dependent on an opioid (δ or μ receptor) synapse in the vlPAG (Tershner and Helmstetter, 1995; Pavlovic, Cooper, and Bodnar, 1996a, 1996b; Pavlovic and Bodnar, 1998a, 1998b). In many respects, many of these pain-signaling systems appear to access directly or indirectly supraspinal pain-inhibition between the vlPAG and the NRM/NRGC. Thus, in all major neuroanatomical aspects, lordosis signaling and pain signaling overlap (Fig. 41).

Furthermore, while most of the information bearing on this argument comes from studies with rats, the neuroanatomical connections of the midbrain PAG in cats support an identical argument for that species (Mouton and Holstege, 1998, 2000).

### b. Descending Pathways

Still more impressive is the overlap of descending pathways related to the control of pain with those governing lordosis behavior. Just as MPOA and VMH outputs exert opposite effects on lordosis behavior (Pfaff, 1980), neurochemical manipulation of these two basal forebrain regions has opposite effects on pain control (Oka et al., 1995).

Perhaps the most obvious registration of reproductive control onto pain control pathways is in the midbrain PAG, especially its ventrolateral quadrant. Here we find estrogen-binding cells (Pfaff, 1973). Further, a large body of data relates PAG neurophysiology to lordosis behavior (Pfaff, 1980), including the stimulating effects of GnRH on reproduction (Sakuma and Pfaff, 1980).

One aspect of understanding the role of the mesencephalic PAG in lordotic and analgesic behaviors is to appreciate that this structure is discretely subdivided anatomically, physiologically, and functionally into dorsal, dorsolateral, lateral, and ventrolateral subdivisions as well as the ventrally oriented cells of the oculomotor complex on its rostral edge, and the DRN on its caudal edge (Beitz, 1982c). Preceding sections have indicated a potential relationship among lordotic, analgesic, and aggressive behaviors, and the latter function has been well characterized in terms of PAG subdivisions. Bandler and Shipley (1994) cogently summarized more than a decade of research in which they proposed a

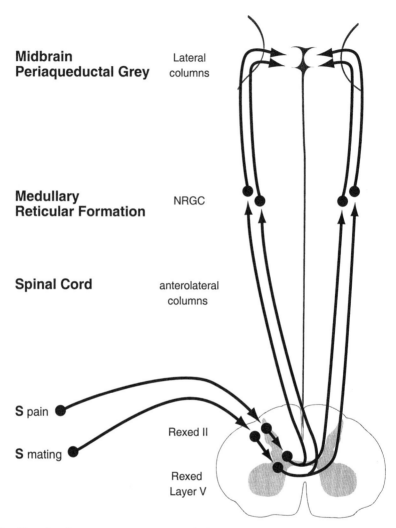

**Fig. 41.** Overlap among neuroanatomical pathways carrying signals ascending in the neuraxis following painful stimuli and cutaneous stimuli relevant for mating behavior is illustrated. The two systems overlap in the dorsal horns of the spinal cord, the ascending spinal cord columns, the medullary reticular formation, and the midbrain PAG. Both systems are partially crossed. S, Stimulus.

columnar organization within the mesencephalic PAG that codes for different aspects of emotional expression. They implicated the vlPAG in the mediation of quiescence, hyporeactivity, hypotension, and bradycardia, as well as its well-known mediation of opioid analgesia following electrical stimulation or opiate microinjections.

In terms of lordotic and aggressive behaviors, activation of the PAG would prevent any confrontational defense or flight reactions, increasing the probability that the female would successfully mate and propogate. Second, PAG activation in all rodent species would promote freezing behavior that is species typical to prevent movement, and thereby "play possum." In many interspecies aggressive encounters, this may stop the predator from inflicting any further damage on the prey, leaving the prey alone, and hence allowing escape. Third, in an end game of an aggressive encounter, PAG stimulation may activate overall analgesic responses that prepare the animal for a fatal blow or death. Bandler and Shipley (1994) implicated the dorsolateral and lateral quadrants of the PAG in defensive behavior that is generally characterized by hypertension, tachycardia, as well as further analgesic responses. The two types of analgesic responses elicited by these two regions can be differentiated. However, the defensive behavior triggered by PAG activation mobilizes the animal toward movement in two critical ways that are subserved differentially by the dorsal and lateral regions of the PAG according to a rostrocaudal delineation.

The rostral portion of the dorsolateral PAG appears to subserve confrontational defense characterized by decreased blood flow to the limbs and viscera and increased blood flow to the face. This posture allows animals to produce affective displays that may chase off potential aggressors. The analgesic response is there to ward off reactions to potential injury sustained by the persistence of the aggressive encounter. This activated area may be of critical importance to female rodents in a nonreceptive state that are approached and encounter sexually charged males. By contrast, the caudal portion of the dorsolateral PAG appears to subserve flight responses that are characterized by increased blood flow to the limbs and viscera and decreased blood flow to the face. This posture maximizes muscular energy for successful escape, and the analgesic response again protects against competing recuperative reactions sustained by potential injury that can interfere with an effective flight-and-escape response. Interestingly, each of these three types of responses (quiescence in the ventrolateral quadrant and either confrontational defense or flight in the dorsolateral quadrant) are mediated locally by microinjections of EAAs that appear intrinsic to each of the three areas. Therefore, it is precisely inputs that descend from hypothalamic, amygdaloid, or other areas that may then modulate each of these subzones into an integrated functional response. We next discuss the relevance of these columnar distinctions within the PAG in further detail for lordotic and analgesic responses.

As indicated earlier, the vlPAG is an extremely important site for SPA. Given the activation of these pathways by dorsal horn nociceptors, it is interesting that tonic descending inhibition of nociceptive dorsal horn neurons is unaffected by lesions of the pontine and medullary raphe nuclei, PAG, and NRGC but is reduced by bilateral lesions of the ventrolateral medulla (Hall et al., 1981, 1982; Duggan and Morton, 1988). By contrast, phasic inhibition of nociceptive dorsal horn responses is observed during SPA. Thus, PAG SPA also produces inhibition of wide dynamic range neurons, yet fails to affect other, nonpain-related cells in lamina IV of the dorsal horn of the spinal cord, in the bulboreticular area of the NRGC, and in the nuclei oralis and caudalis or the trigeminal nerve (Liebeskind et al., 1973; Oliveras et al., 1974; LeBars et al., 1975; Fields and Anderson, 1978). vlPAG SPA respectively excites the NRM and inhibits laminae I, II, and V of the dorsal horn (Oliveras et al., 1974; Behbehani and Pomeroy, 1978; Fields and Anderson, 1978; Behbehani and Fields, 1979). Similarly, NRM stimulation produces analgesia, and inhibition of nociceptive dorsal horn neurons that were characterized as either wide-range dynamic neurons or identified nociceptors in lamina I (Gebhart et al., 1983a, 1983b). The pathway through which such phasic inhibition from the NRM to the dorsal horn occurs is either the ipsilateral (Fields and Basbaum, 1978) or contralateral (Sandkuhler, Maisch, and Zimmermann, 1987) dorsolateral funiculus. Moreover, identified pathways from the RVM to catecholamine (A5, A6, A7) nuclei provide another source of descending, adrenergic inhibition (Proudfit, 1988, 1992; Proudfit and Yeomans, 1995). In addition to inhibiting the dorsal horn nociceptors, stimulation of the PAG or NRM inhibits identified cells in the spinothalamic tract that are of both the wide dynamic range and nociceptive-specific varieties (Willis, 1985, 1988). Nor is this argument limited to small laboratory rodents: the neuroanatomical analysis of descending connections from the PAG of the cat supports this argument as well (Mouton and Holstege, 1994; Mouton, Kerstens, and Holstege, 1996).

Note that SPA can be elicited from other more dorsal subdivisions of the vlPAG that can be clearly differentiated from vlPAG SPA. Whereas vlPAG SPA is accompanied by quiescence and freezing, stimulation of the dorsolateral quadrant is accompanied by explosive motor behavior and escape (Fardin, Oliveras, and Besson, 1984a, 1984b; Oliveras and Besson, 1988). PAG SPA has been associated with opiate systems because of its ability to display tolerance following repeated administration, and cross-tolerance with morphine (Mayer and Hayes, 1975). Lesions placed in the dorsolateral funiculus of the spinal cord block SPA as well as morphine analgesia (Basbaum et al., 1977). PAG SPA has opioid and nonopioid components, an interpretation based on its sensitivity and insensitivity to naloxone since SPA elicited from dorsal PAG sites was insensitive to opiate antagonism, and SPA elicited from ventral PAG sites, including the DRN, was sensitive to opiate antagonism (Cannon et al., 1982) and other opioid manipulations (Nichols, Thorn, and Berntson, 1989) at spinal and supraspinal levels (Morgan, John, and Liebeskind, 1989). NRM lesions differentiate between SPA elicited from dorsal and ventral PAG sites, with reductions noted at the latter but not former sites (Prieto, Cannon, and Liebeskind 1983). By contrast, analgesia elicited by dorsal PAG stimulation is blocked by lesions placed in the ventrolateral medullary area (Lovick, 1985). These opioid forms of PAG SPA appear to interact specifically with opioid forms of stress-induced analgesia such that opioid-mediated, prolonged, intermittent foot-shock analgesia displays cross-tolerance with SPA elicited from the ventral but not the dorsal aspect of the PAG (Terman, Penner, and Liebeskind, 1985) and is correlated in genetic studies breeding animals for high and low levels of stress-induced analgesia (Marek et al., 1989; Marek, Yirmiya, and Liebeskind, 1990).

For lordosis behavior, midbrain PAG fibers descend to the level of the medulla, where they help activate medullary reticulospinal neurons, especially on the ventral medial side of the medullary reticular formation. Likewise, the RVM is important for SPA (Oliveras et al., 1974, 1978). Both electrical stimulation of and morphine microinjection into the vlPAG respectively excite the NRM and inhibit laminae I, II, and V of the dorsal horn (Oliveras et al., 1974; Behbehani and Pomeroy, 1978; Behbehani and Fields, 1979).

Recently, a great deal of more specific information has emerged in which pain control mediated by RVM neurons supports our concept of an overlap between SPA pathways and reproductive behavior mechanisms. RVM microstimulation in lightly anesthetized rats suppresses the tail-flick reflex (Hentall, Barbaro, and Fields, 1991) as part of a system linking SPA mechanisms in the midbrain PAG and the spinal cord (Fields, Heinricher, and Mason, 1991; Morgan, Heinricher, and Fields, 1992). As outlined previously, this system depends on the considerable densities of opioid-containing neurons found in the midbrain PAG and RVM (Mansour et al., 1988, 1995) as well as the two types of RVM cells (off-cells and on-cells) that modulate nociceptive responses and predict systemic morphine analgesia. When PAG neurons produce antinoci-

ception, endogenous opioid peptides are released in the RVM and selectively inhibit those on-cells that would have otherwise had a facilitating action on spinal nociceptive transmission (Pan and Fields, 1996). In
fact, morphine applied either intrathecally or in the PAG exerts similar
effects on RVM neurons (Heinricher, Haws, and Fields, 1991). Likewise,
opioid actions in the RVM disinhibit off-cells, thereby allowing expression of analgesic effects (Heinricher et al., 1994). The importance of
such RVM neurons to pain control is indicated by the lack of antinociception following inactivation of the RVM by local anesthesia (lidocaine) or GABA agonists (muscimol) (Mitchell, Lowe, and Fields, 1998).
Quite separate from serotonergic neurons in the medullary raphe, nonserotonergic magnocellular reticular neurons have response characteristics suggesting that both on-cells and off-cells are likely to modulate responses to subsequent noxious insults (Leung and Mason, 1998). The
contributions of serotonergic neurons have been previously emphasized
in neuroanatomical and neuropharmacological studies, including our
observations that RVM administration of general, 5-HT-2, and 5-HT-3 receptor antagonists blocks vlPAG morphine analgesia (Kiefel, Cooper,
and Bodnar, 1992). In fact, serotonergic terminals are found on both
nonserotonergic and other serotonergic cells, with the latter cells physiologically characterized as "neutral cells" (Potrebic, Fields, and Mason,
1994; Potrebic, Mason, and Fields, 1995), which fail to predict the analgesic effects of either vlPAG electrical stimulation or morphine (Gao,
Kim, and Mason, 1997; Mason, 1997; Gao et al., 1998).

By contrast, EAA transmitters have been shown to be important for
opioid-sensitive analgesic actions in the RVM such that RVM NMDA antagonists significantly reduce vlPAG morphine analgesia (vanPraag and
Frenk, 1990; Spinella, Cooper, and Bodnar, 1996). Further, EAA transmission appears to be responsible for RVM on-cell and off-cell activation
(Heinricher and Roychowdhury, 1997; Heinricher and McGaraughty,
1998) such that RVM EAA antagonism blocked morphine analgesia and
the opioid-induced activation of off-cells (Heinricher, McGaraughty, and
Farr, 1999). Moreover, blockade of RVM GABA transmission in the RVM
is implicated in analgesic responses by both electrophysiological and behavioral evidence (Heinricher et al., 1994; Pan and Fields, 1996; Roychowdhury and Fields, 1996). Because descending lordosis pathways
from the midbrain PAG terminate in this same region of the medulla,
the great richness of pain control phenomena deriving from experiments with these medullary neurons gives further evidence of an overlap between reproductive behavior and antinociceptive mechanisms in

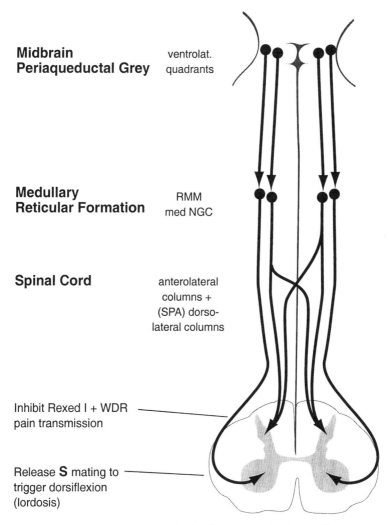

**Midbrain**      ventrolat.
**Periaqueductal Grey**    quadrants

**Medullary**      RMM
**Reticular Formation**    med NGC

**Spinal Cord**      anterolateral
columns +
(SPA) dorso-
lateral columns

Inhibit Rexed I + WDR
pain transmission

Release **S** mating to
trigger dorsiflexion
(lordosis)

**Fig. 42.** Overlap among neuroanatomical pathways carrying descending motor control information relevant to stimulus-produced analgesia and female-typical mating behavior is illustrated. The two systems overlap in the ventrolateral quadrants of the midbrain PAG and the medullary reticular formation. The two systems clearly diverge at levels of the spinal cord where the two separate response systems are controlled. WDR, Wide dynamic range; S, stimulus.

the medulla. Since their projections will diverge as they descend into the spinal cord, the medulla may be another and perhaps the final site at which interactions between these two behaviors take place.

As lordosis-controlling and analgesia-causing signals descend from medial reticulospinal neurons and travel through and into the spinal cord, they begin to diverge, as one would expect from the simple fact that the two response patterns are not identical. Lordosis-relevant descending influences travel through the anterolateral column of the spinal cord, whereas most axons related to SPA appear to travel through the dorsolateral columns (Basbaum et al., 1977; Rhodes, 1979), and depend on serotonergic and noradrenergic terminals in the spinal cord for the full expression of PAG SPA (Akil and Mayer, 1972; Akil and Liebeskind, 1975; Yaksh and Rudy, 1978).

Thus, there is a high degree of overlap between neuronal groups subserving female-typical reproductive behaviors and SPA, all the way from the MPOA and VMH back through the posterior medulla (Fig. 42). The similarity between these mechanisms raises the strong possibility that one aspect of reproductive behavior facilitation is to control responses to noxious inputs by activating these descending centrifugal systems.

## H. Summary

Against a background of evidence demonstrating opioid peptides and their receptors controlling pain, a connection to reproduction is obvious. An important genomic effect of estrogen in VMH neurons is to turn on the enkephalin gene. This genetic induction and its consequences activate descending analgesic pathways that, to a large extent, are coextensive with lordosis behavior pathways. Such a descending system renders somatosensory and visceral stimuli that ordinarily would be extraordinarily noxious, much more tolerable, so that the male can mount and fertilize the female when she performs mating behaviors.

# IV

## Inferences and Arguments

Through the understanding of hormonal effects on ascending arousal systems (covered in chapter II), coupled with the apparent overlap of lordotic behavior and *analgesia-producing circuitry* (reviewed in chapter III), we see that estrogens followed by progesterone could produce a *sexually aroused* female willing to tolerate *strong somatosensory stimuli* from a male. Therefore, this physiological combination adds two concepts to the basic working circuit for producing lordosis behavior, which has been spelled out in neuroanatomical, neurophysiological, and molecular detail (Pfaff, 1980, 1999a). The precision, reliability, and mechanistic detail in which the neural operations and hormonal effects underlying this mammalian social behavior have been elucidated all help to correct ancient prejudices against behavioral research. It also used to be thought that behavioral studies were not quantitative and had little to do with the practice of medicine.

Regarding precision, first, the old point of view failed to take account of the fact that many students entering behavioral science had backgrounds in their studies of physics and mathematics. An amusing way of recounting a high degree of reliability inherent in certain aspects of behavioral research would compare historical achievements in this regard to those in genetics. Considering that modern genetics had its inception around the year 1900, it is gratifying to recall the laws of psychophysics, Fechner's among them, discovered during the 1880s, which could be replicated in quantitative detail even today, more than a century later. Second, in 1944 at the Rockefeller University, DNA was discovered and announced as the genetic substance. Yet, several years before, in 1938, quantitative laws governing contingencies of reinforcement were published by the great psychologist B. F. Skinner in "The Behavior of Organisms." To this day, precise control over stimuli, reliable measures of

165

responses, careful experimental design, and incisive mathematical and statistical analysis of data remain hallmarks of good behavioral research.

Regarding medical importance, several of the leading causes of death in the United States and a majority of the most important causes of disabilities worldwide have behavioral roots. Autoimmune deficiency syndrome, drug addiction, diseases due to smoking cigarettes, certain forms of heart disease, suicide, and deaths resulting from uncontrolled aggression all represent obvious examples. What could be more medically important than these? Moreover, preventive medicine is not only cost-effective in societal terms, but also allows individuals and families to deal proactively with health risks and thus save enormous grief. Clearly, preventive medicine requires understanding the patterns of human behavior.

The great twentieth-century neurophysiologist Vernon B. Mountcastle, in his article "Brain Science at the Century's Ebb" (cited in Pfaff, 1999a), quoted the philosopher Isaiah Berlin as follows: "The ideal of all natural science is a system of propositions so general, so clear, so comprehensive, connected with each other by logical links so unambiguous and direct that the result resembles as closely as possible a deductive system, where one can travel along wholly reliable routes from any point on the system to any other." Neither Professor Mountcastle nor anyone else imagines that neuroscience has reached this stage of development. Nevertheless, it is surprising that hormonal effects on certain mammalian social behaviors, manifest as part of the science of neuroendocrinology, have allowed comprehensive and reliable analyses of causes and mechanisms (Pfaff, 1999a). Added to that edifice by the data and arguments in this book are two aspects summarized next. In section A, proofs of causal gene/behavior relationships are followed by analyses of causal routes. Most important, the specificity with which these causal routes operate demands that the development of behavioral science keep pace with the elaboration of genetic studies. In section B, a "formula" for lordosis behavior, which depends on elevated levels of both arousal and analgesia, links pain to sex in a manner that is understandable neuroanatomically and potentially important for psychiatry.

## A. Gene/Behavior Relationships: Application to Opioid Peptides

Thinking ahead toward the great expanse of behavioral research forthcoming, to elucidate gene/behavior relationships in most cases will be extremely difficult, as recently illustrated (Crabbe, Wahlsten, and Dudek, 1999). First, basic genetics: The pleiotropy of gene action (the ability of a gene to contribute to more than one function, especially when different stages of the life cycle are considered), the redundancy of genetic effects (overlapping functions of genes that allow one to substitute for another), and our lack of understanding of the quantitative degree of penetrance of dominant alleles in heterozygotes (rendering it difficult to construct gene dose/response relationships) all will throw down the gauntlet to the aspiring geneticist wishing to understand behavior (Fig. 43) (reviewed in Pfaff, 1999b). Second, the very specificity of gene/behavior causal effects, illustrated fivefold later, will make it impossible to construct simple lists of genes and behaviors and draw one-to-one connections between them as might have been expected by the simplest cases in Drosophila (Fig. 44). Therefore, it will not be feasible to do, as Drosophila geneticists do, our reasoning "from the gene on out" to a pattern of behavior. That type of reasoning might be required for disease genes in human maladies, in which the mutation constitutes the entirety of the offending item. However, for a genetic analysis of the normal physiology of behavior, animal or human, understanding the biological function of the behavior and the systematic properties of its neuroanatomical, neurophysiological, and neurochemical mechanisms will be required before there can be a full appreciation of how genetic contributions have their effects during development or in adulthood (Ogawa et al., 1996a).

Nevertheless, in the clearest cases, it has been possible to prove a causal relationship between an individual mammalian gene and a particular mammalian social behavior. The classical ER gene is the first example, and its product is absolutely required for the performance of lordosis behavior in female mice (Figs. 45 and 46) (Ogawa et al., 1996a, 1996b). Most interesting is the fact that a newly discovered ER gene, ER-β, is not at all necessary for the performance of lordosis (Ogawa et al., 1999) (Fig. 47). Even more intriguing is that loss of the classical ER gene tends to masculinize a range of behaviors in genetic females, yet tends to feminize a range of behaviors in genetic males (Fig. 48) (Ogawa et al., 1997, 1998a, 1998b).

In two cases, the richness of data in the molecular biology of behavioral neuroendocrinology have allowed us to construct the *causal routes* by which the two genes act. First, the classical ER gene, ER-α, acts as a transcription factor in hypothalamic neurons to foster the synthesis of at least six products that contribute to different aspects of female reproductive behavior (see review in Pfaff [1999a]). The contributions of different gene products have different biological roles, as illustrated sub-

# A

## PLEIOTROPY

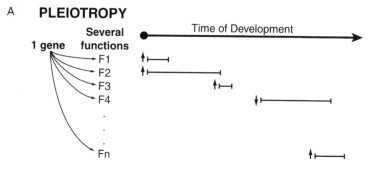

## REDUNDANCY

## VARIATIONS IN PENETRANCE OF DOMINANT ALLELES IN HETEROZYGOTES OF DIFFERENT GENES

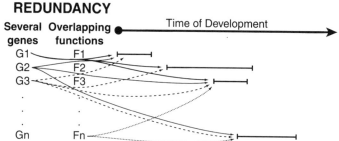

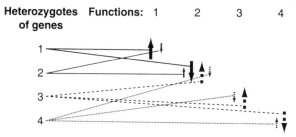

# B

TIME

| | | Through Development | | During Adulthood | |
|---|---|---|---|---|---|
| | | Lethal | Not Lethal | Sensory/Motor | Integrative |
| **L** **O** **C** | DIRECT ON BRAIN | | | | |
| **U** **S** | ACTION OUTSIDE BRAIN | | | | |

INDIRECT

DIRECT

sequently for opioid peptides and their receptor genes. Second, the gene for thyroid hormone receptor-β acts to block the transcriptional effects of ER-α, thus reducing reproductive behavior, perhaps to prevent inappropriate reproduction under unforgiving metabolic or environmental conditions. It is fascinating that the effects of two genes coding for thyroid hormone receptors, TR-β and TR-α, have opposite effects on lordosis behavior (Dellovade et al., 2000).

The specificity of gene/behavior causal relationships in mammals, as we have studied them in the arena of natural behaviors, is paramount. First, the effect of a gene on a mammalian social behavior can depend on the *gender* of the animal in which that gene is expressed: the ER gene disruption masculinizes the behavior of females but feminizes the behavior of males (Fig. 48) (Ogawa et al., 1996a, 1996b, 1997, 1998a, 1998b). Second, the effect of a gene on a mammalian behavior depends on precisely *when and where* that gene is expressed: temporary interruption of ER gene expression specifically during the neonatal period in the hypothalamus (McCarthy, Schlenker, and Pfaff, 1993) does not have the same effect on lordosis behavior as a permanent, bodywide disruption of the ER gene in female mice. Third, the effect of a gene on complex natural behaviors such as parental behavior can depend on the level of *ambient environmental stress* (Pfaff, 1999b; Ragnauth et al., 2001). Fourth, the effect of a gene disruption on aggressive behavior can depend on the *gender of the opponent* (Ogawa et al., unpublished observations). Fifth, with tests of aggression in ER-β knockout mice, the quantitative comparison between the knockouts and their wild-type controls depended on the *degree of previous social experience with aggression* (Ogawa et al., 1999). All of these data tell us that comprehensive investigation of the lawfulness and biological appropriateness of a given behavior will be required to understand the existence, direction, nature, and causal routes of the contributions from any particular set of genes.

The advanced development of mechanistic studies in reproductive behavior has allowed us to solve the question of how genes (**G**), internal stimuli (**In**), and external stimuli (**Ex**) interact to cause behavior. Only in the presence of circulating estrogens (**In**) and, to an even greater ex-

**Fig. 43.** Illustrations of potential difficulties in discerning gene/behavior relations in mammals. (*A*) Fundamental genetic reasons are shown: pleiotropy of gene action, overlapping genetic functions, and our lack of understanding of penetrance. (*B*) While many neurobiologists want most to understand the effects of gene expression on integrative actions of the nervous system during adulthood ("direct effects"), a multitude of indirect routes of genetic influences also can easily be seen.

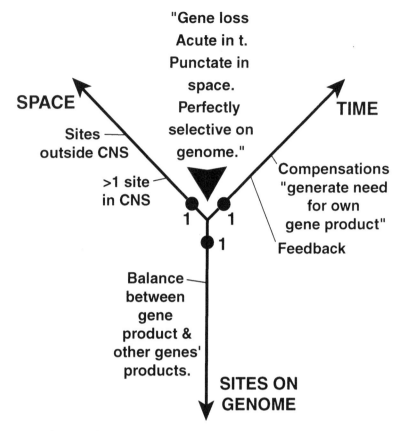

**Fig. 44.**   Theoretical mapping of a "3-space" to illustrate the difference between the interpretation of gene knockout effects often desired by mammalian neurobiologists compared to the large number of possibilities available. The axes in the "3-space" are space, time, and sites on the genome. While we often wish that the effects of gene loss were acute in time, effective only at the neurons involved in the function under study, and with no further implications for genomic structure (coordinates in the theoretical 3-space = 1, 1, 1,), such is never really the case. Regarding space, the gene loss may be effective elsewhere in the CNS and even at sites outside the CNS. Regarding time, a gene knockout allows time for compensations and for a variety of feedback loops to operate. Regarding the genome, there are obvious possibilities for changes in genomic structure and interactions with other gene products subsequent to an individual gene's knockout.

tent, when estrogens are amplified by progestins (**In**), somatosensory stimuli from the male (**Ex**) will trigger lordosis behavior in the genetic female (**G**). The enormous sex difference in this behavior shows that an alteration of genotype by adding a Y chromosome (**G**) blocks the behavior. Moreover, in a developmental context, this interactionist type of thinking (**G** × **In** × **Ex**) subsumes nature/nurture controversies. Thus, only the proper combination of genotype, internal stimulus, and external stimulus permits lordosis behavior.

In a broader context (Fig. 49), hormonal and sensory stimulus effects are integrated in the presence of permissive environmental stimuli (adequate nutrition, adequate water, adequate salt supply, nest material pres-

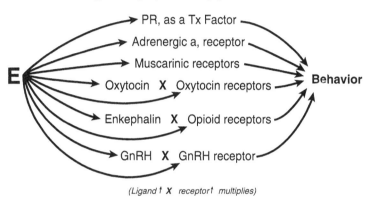

**A**

## Gene Products Elevated by Estrogen (E) in Hypothalamus

E → PR, as a Tx Factor → Behavior
→ Adrenergic a, receptor
→ Muscarinic receptors
→ Oxytocin **X** Oxytocin receptors
→ Enkephalin **X** Opioid receptors
→ GnRH **X** GnRH receptor

*(Ligand ↑ X receptor↑ multiplies)*

| B | Incidence of female reproductive behavior | Incidence of female - female aggression |
|---|---|---|
| Wild type | normal | 2/21 mice |
| Estrogen Receptor KnockOut | none | 10/25 mice* |

**Fig. 45.** Relations between gene expression and lordosis behavior. (*A*) Estradiol (E) acting through ERs turns on a variety of genes with implications for lordosis behavior. This list has been made longer in our laboratory through the use of differential display, and in a wider-scope search for new gene products, DNA chip/microarray techniques currently are being employed. (*B*) Under testing circumstances in which 100 percent of the wild-type female mice would show lordosis behavior, 0 percent of the ER gene knockout females (ERKO) showed lordosis. * = the form of aggression exhibited by ERKO females comprised mainly offensive attacks more typical of intermale aggression. (From Ogawa et al., 1996a, 1996b.)

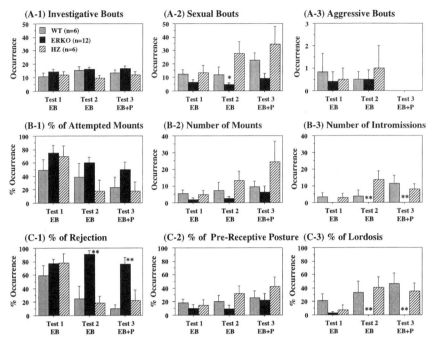

**Fig. 46.** The results of further studies with ER gene knockout female mice (ERKO) under circumstances in which we controlled hormone levels are shown. All females were OVX and given replacement hormone therapy consisting of estradiol benzoate (EB) or estradiol benzoate plus progesterone (EB+P). Even under circumstances in which ERKO females received social investigation (A-1) and sexual investigation (A-2) as well as attempted mounts (B-1) and actual mounts (B-2), the ERKO females displayed little or no lordosis behavior (C-3). As a result, the males were never able to achieve intromissions (B-3). This did not result because the ERKO females were engaged in frank aggressive behavior (A-3), but they did increasingly reject mount attempts by the male (C-1). (From Ogawa et al., 1998a.) WT, Wild-type; HZ, heterozygotes. *, Significant behavioral alterations in ER knockout animals relative to both wild-type controls and their heterozygous littermates.

ent, low ambient stress levels, permissive temperature, sufficient length of day). This integration produces temporally coordinated endocrine outputs and behavioral outputs: lordosis behavior permits fertilization around the time of the ovulatory LH surge. Autonomic changes feed in, to support the behavioral activities—muscular efforts—required for reproductive behaviors.

Since, as reviewed in chapter III, section B, estrogens turn on the enkephalin gene in hypothalamic neurons and, in addition, increase the levels of mRNAs for δ and μ opioid receptors through which enkephalin acts, what special points can we make about gene/behavior relationships in mammals using the example of opioid peptides? First, because the synthetic capacity for the ligand, enkephalins, and for its receptors are increased by estrogens, there is the possibility of a *multiplicative* action of steroid sex hormones. Second, from the recent results of Andre Ragnauth and colleagues at Rockefeller University and at Queens College, we understand that the behavioral impact of the enkephalin gene might be magnified by higher levels of ambient stress. Importantly, the action of the enkephalin gene, studied as a gene knockout prepared by Professor John Pintar at Rutgers University, could protect against fear-laden disruptions of natural behavioral sequences (Ragnauth et al., 2001). Most exciting is the fact that estrogen turns on the enkephalin gene very rapidly, long before lordosis behavior will occur, and thus is likely to allow the female to locomote and explore in the manner of courtship behaviors, advertising her sexual receptivity through the use of pheromones. Finally, the action of an opioid peptide, operating through hormone-

### COMPARISONS of ER-$\alpha$ & ER-$\beta$ FUNCTIONS in CNS

| Necessary? | Sufficient? | | ASSAYS IN THE ♀ | ASSAYS IN THE ♂ |
|---|---|---|---|---|
| ER $\alpha$&$\beta$ | **Neither** $\alpha$ **nor** $\beta$ | "ER $\alpha$&$\beta$ must synergize." | none | none |
| Neither $\alpha$ nor $\beta$ | **Either** $\alpha$ **or** $\beta$ | "ER $\alpha$&$\beta$ can subst for each other." | E induction of PR (ICC) E reduction of ER (ICC) | E induction of PR (ICC) E reduction of ER (ICC) |
| $\alpha$, $\beta$ each for it's own | $\alpha$, $\beta$ **each for it's own** | "ER $\alpha$ vs $\beta$ contribs. not related." | Maternal behavior ($\alpha$) Suppression of aggression ($\alpha$) Reduction of food intake ($\alpha$) Reduction of anxiety ($\alpha$) | Mounting behavior ($\alpha$) Anxiety response ($\alpha$) |
| ER $\alpha$ absent $\beta$ | **ER $\alpha$ absent $\beta$** | "ER $\beta$ can reduce $\alpha$ effect." | Lordosis behavior | Intromissions & ejaculation Aggression |
| Optimal $\alpha$/$\beta$ balance | **Optimal $\alpha$/$\beta$ balance** | "ER $\alpha$ vs $\beta$ always opposed." | none | none |

**Fig. 47.** Scenarios for possible relations between ER-$\alpha$ and ER-$\beta$ functions in the CNS are plotted on the left. Results of histochemical and behavioral assays conducted in the genetic female and the genetic male are plotted on the right. Note that the extreme circumstances never obtain. That is, we have no data indicating that ER-$\alpha$ and ER-$\beta$ must synergize, nor do we have data showing that ER-$\alpha$ and ER-$\beta$ functions are equal and opposite. From this and other work in our laboratory, it seems clear that modern molecular neurobiologists must struggle to take a step beyond the classical Beadle and Tatum formulation, the "one gene, one enzyme" hypothesis. In other words, we will be looking primarily for *patterns* of neural and behavioral functions altered by *patterns* of gene expression. ICC, Immunocytochemistry.

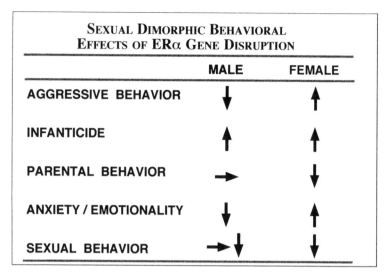

**Fig. 48.** Effects of the knockout of the classical ER gene, ER-α, are not the same between genetic males and females.

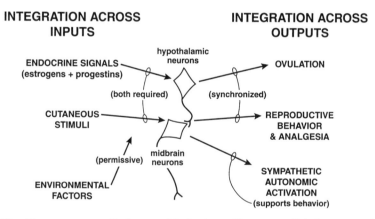

**Fig. 49.** Hormone-controlled natural behaviors offer unparalleled opportunities to study the mechanisms of integration among neural, behavioral, and other physiological influences. On the input side, endocrine signals and specific cutaneous stimuli are both required for normal female reproductive behaviors to occur. Environmental factors such as availability of food, temperature, and time of day play a permissive role. On the output side, the timing of ovulation and reproductive behavior is virtually synchronized in a biologically sensible fashion. Arousal of the female is correlated with activation of the sympathetic autonomic nervous system in a manner that supports vigorous behavioral activity.

enhanced opioid receptors, is likely to add the analgesia necessary for the female to tolerate cutaneous stimuli from the male. In this instance, we understand that gene products for neuropeptides do not simply "add up" to comprise a behavioral mechanism, but, instead, they participate by supplying particular signals and performing particular biological functions. Nor are the clear examples limited to reproduction. Strain differences, gender differences, and even individual differences in sensitivity to pain have been laid to genetic influences (Mogil, 1999).

What do all these findings mean for genetic influences on the behavior of humans? With all the caveats we have noted, genetic contributions to the understanding of certain mental illnesses have been reviewed (Pfaff, 1999b). In the context of this book's subject matter, it is fascinating that the human behavioral phenomena surrounding behavioral inhibition, fear, shyness, and social phobias (Schmidt and Schulkin, 1999) would appear to be congruent with the trait called "harm avoidance," whose chromosomal origins are among the best documented (Svrakic, Przybeck, and Cloninger, 1992; Svrakic et al., 1993; Stallings and Hewitt, 1996).

## B. Biological Importance of the Relations among Sex, Arousal, and Analgesia

In this book, we have elaborated the neurophysiology and neurochemistry underlying two concepts: arousal and analgesia. These embellish the basic working neural circuit and basic molecular genetics for lordosis behavior that have already been worked out and reported (Pfaff, 1999a). First, we have the concept that steroid sex hormones raise the general arousal state of a female, resulting in, among other things, heightened sexual arousal; and, second, the concept that hormone-facilitated neural and genetic systems permit a form of analgesia. These two concepts lead to a new kind of "formula" for this type of mating behavior. In the basic working female reproductive behavior circuit, "Arousal + Analgesia → lordosis behavior."

Estrogen causes a change in CNS drive state, which, for the purpose of controlling reproduction, has at least two components, which, in turn, have an intricate relation to each other. First, estrogen *arouses* the female and affects preoptic neurons to cause increased locomotor and courtship behavior. The female is alert and muscularly taut, prepared to support the weight of the male during mating behavior. Second, through genomic and electrophysiological actions in VMH neurons, estrogen causes a specific form of *analgesia* that permits lordosis behavior by rendering strong stimuli from the male more tolerable. The normal opposition between MPOA and VMH mechanisms, manifest in the requirement for the female to stop locomoting in order to accomplish lordosis, guarantees that the two forms of behavior cannot occur at the same time—they must occur sequentially: courtship behavior followed by lordosis, followed in turn by the next episode of locomotion, followed again by lordosis, and so forth. Thus, genomic analysis and neurophysiology match the normal behavioral sequence precisely.

What are the chances that sexual motivation, general arousal, and mood states all boil down essentially to the same thing or, at least, that they share important component mechanisms? After all, the same five neurochemical systems reviewed in chapter II participate in the generation of broadly defined neural states. The easiest concepts would have emphasized those aspects of the mechanisms of motivation, mood, and arousal, which in their component mechanisms are truly generalized (Table 1). For both sex and pain, strong cutaneous stimuli leading to broadly distributed electrophysiological responses in the dorsal horn of the spinal cord and in the brain stem reticular formation will encounter neurochemical systems whose outputs are not precise in a point-to-point fashion but instead are distributed widely (Table 1). Likewise, it is already known that the functional effects of these neurochemical systems are not limited to particular stimulus-response combinations. Motivational theorists such as Hebb, Hull, and Duffy all called our attention to important generalized motivational components, based on behavioral evidence (Fig. 9), as did classical neurophysiological researchers who have dealt with sleep, vegetative states, and coma. In the classification of mood states, as well, general arousal plays an obvious role, since "mood maps" generated by Plutchik and others use at least two dimensions: pleasantness and, more important, intensity. On all of these grounds, one expects considerable overlaps between the phenomena of pain and sex. In the animal brain, such overlap guides our study of neural and genetic mechanisms. The data from Seymour Antelman and colleagues, showing activation of male rat mating behavior by tail pinch, were among the first to relate the two arenas of thought directly.

On the other hand, there are multiple opportunities for specificity. In particular, we must retain a view of how specific reproductive behaviors can be motivated. Indeed, in the normal case, arousal due to sex arises from different stimuli than for arousal due to pain, and reflects a different shading of motivation using neuroanatomical pathways that are not identical and neurochemical systems that are not identical, resulting in different behavioral responses (Table 1). We must conclude, therefore, that the arousal/motivational systems underlying sex and pain overlap considerably but are discrete.

In summary, estrogen causes a change in CNS state, yielding a female alert, muscularly taut, active, satiated (guaranteed by glucose and other nutritional effects on ventromedial hypothalamic neurons—the female should not be reproducing unless there is adequate nutrition), tolerant of somatosensory stimulation, and ready for reproduction. The combination of a high state of arousal with a selective analgesia caused by estrogenic actions in the forebrain and midbrain yields descending systems that *permit* spinal cord management of lordosis behavior.

Besides proving that it is thus possible to bridge the gulf between genomic and neurophysiological mechanisms, on the one hand, and behavioral realities, on the other, these data and inferences have an important implication. They show how extremely general arousal systems and broadly gauged analgesia systems for antinociception can foster very specific and timely behavioral responses, whose performance, in turn, reflects the change in central motivational state. At one stroke, therefore, elucidating these neurophysiological states both enriches our understanding of specific mammalian sociosexual behaviors whose neural and genetic mechanisms already have been discerned (Pfaff, 1999a) and reveals systems components shared by these sociosexual mechanisms with neural pathways subserving analgesia.

# References

Abbott, F.V., and Palmour, R.M. 1988. Morphine-6-glucuronide: analgesic effects and receptor binding profile in rats. *Life Sci* 43:1685–95.

Abercrombie, E., and Jacobs, B. 1987a. Single-unit response of noradrenergic neurons in the locus coeruleus of freely moving cats. I. Acutely presented stressful stimuli. *J Neurosci* 7:2837–43.

———. 1987b. Single-unit response of noradrenergic neurons in the locus coeruleus of freely moving cats. II. Adaptation to chronically presented stressful stimuli. *J Neurosci* 7:2844–48.

Abols, I.A., and Basbaum, A.I. 1981. Afferent connections of the rostral medulla of the cat: a neural substrate for midbrain-medullary interactions. *J Comp Neurol* 201:285–97.

Adams, J.E. 1976. Naloxone reversal of analgesia produced by brain stimulation in the human. *Pain* 2:161–66.

Adams, J.U., Chen, X., DeRiel, J.K., Adler, M.W., and Liu-Chen, L.Y. 1994. Intracerebroventricular treatment with an antisense oligodeoxynucleotide to κ-opioid receptors inhibited κ-agonist-induced analgesia in rats. *Brain Res* 667:129–32.

Agmo, A., and Gomez, M. 1993. Sexual reinforcement is blocked by infusion of naloxone into the medial preoptic area. *Behav Neurosci* 107(5):812–18.

Aimone, L.D., and Gebhart, G.F. 1986. Stimulation-produced spinal inhibition from the midbrain in the rat is mediated by an excitatory amino acid transmitter in the medial medulla. *J Neurosci* 6:1803–13.

Aimone, L.D., Jones, S.L., and Gebhart, G.F. 1987. Stimulation-produced descending inhibition from the periaqueductal gray and nucleus raphe magnus in the rat: mediation by spinal monoamines, but not opioids. *Pain* 31:123–36.

Akaike, A., Shibata, T., Satoh, M., and Takagi, H. 1978. Analgesia induced by microinjection of morphine into, and electrical stimulation of, the nucleus reticularis paragigantocellularis of rat medulla oblongata. *Neuropharmacology* 17:775–78.

Akesson, T.R., and Micevych, P.E. 1991. Endogenous opioid-immunoreactive neurons of the ventromedial hypothalamic nucleus concentrate estrogen in male and female rats. *J Neurosci Res* 28:359–66.

Akesson, T., Simerly, R., and Micevych, P. 1988. Estrogen-concentrating hypo-

179

thalamic and limbic neurons project to the medial preoptic nucleus. *Brain Res* 451:381–85.

Akil, H., and Liebeskind, J.C. 1975. Monoaminergic mechanisms of stimulation-produced analgesia. *Brain Res* 94:279–96.

Akil, H., and Mayer, D.J. 1972. Antagonism of stimulation-produced analgesia by PCPA. *Brain Res* 44:692–97.

Akil, H., Madden, J., Patrick, R.L., and Barchas, J.D. 1976. Stress-induced increase in endogenous opioid peptides: concurrent analgesia and its partial reversal by naloxone. In *Opiates and Endogenous Opioid Peptides,* ed. H. W. Kosterlitz, 63–70. Amsterdam: Elsevier.

Akil, H., Watson, S.J., Young, E., Lewis, M.E., Khachaturian, H., and Walker, J.M. 1984. Endogenous opioids: biology and function. *Annu Rev Neurosci* 7:223–55.

Alleva, E., Castellano, C., and Oliverio, A. 1980. Effects of l-and d-amino acids on analgesia and locomotor activity of mice: their interaction with morphine. *Brain Res* 198:249–52.

Al-Rodhan, N., Chipkin, R., and Yaksh, T.L. 1990. The antinociceptive effects of SCH-32615, a neutral endopeptidase (enkephalinase) inhibitor, microinjected into the periaqueductal gray, ventral medulla and amygdala. *Brain Res* 520:123–30.

Altier, N., and Stewart, J. 1999. The role of dopamine in the nucleus accumbens in analgesia. *Life Sci* 65:2269–87.

Alves, S.E., Weiland, N.G., Hayashi, J., and McEwen, B.S. 1998. Immunocytochemical localization of nuclear estrogen receptors and progestin receptors within the rat dorsal raphe nucleus. *J Comp Neurol* 391:322–34.

Alves, S., McEwen, B., Hayashi, S., Korach, K., Pfaff, D., and Ogawa, S. 2001. Estrogen regulated progestin receptors are found in the midbrain raphe but not hippocampus of estrogen receptor α (ERα) gene disrupted mice. *J Comp Neurol.* In press.

Alves, S., McEwen, B., Korach, K., Pfaff, D., and Ogawa, S. 1999. Estrogen-regulated progestin receptors are found in serotonin cells in the ERαKO mouse. *Soc Neurosci Abstr* 581(2):1448.

Amaral, D.G., and Campbell, M.J. 1986. Transmitter systems in the primate dentate gyrus. *Hum Neurobiol* 5(3):169–80.

Amir, S., Brown, Z.W., and Amit, Z. 1980. The role of endorphins in stress: evidence and speculations. *Neurosci Biobehav Rev* 4:77–86.

Ansonoff, M., and Etgen, A. 1998. Estradiol elevates protein kinase C catalytic activity in the preoptic area of female rats. *Endocrinology* 139:3050–56.

Anton, B., Fein, J., To, T., Li, X., Silberstein, L., and Evans, C.J. 1996. Immunohistochemical localization of ORL-1 in the central nervous system of the rat. *J Comp Neurol* 368:229–51.

Appelbaum, B.D., and Holtzman, S.G. 1984. Characterization of stress-induced potentiation of opioid effects in the rat. *J Pharmacol Exp Ther* 231:555–65.

————. 1985. Restraint stress enhances morphine-induced analgesia in the rat without changing apparent affinity of receptor. *Life Sci* 36:1069–74.

Arnsten, A.F.T. 1998. The biology of being frazzled. *Science* 280:1711–12.

Arpels, J. 1996. The female brain hypoestrogenic continuum from the premenstrual syndrome to menopause. *Reprod Med* 41:633–39.

Arvidsson, U., Dado, R., Reidl, M., Lee, J.-H., Law, P.-Y., Loh, H., and Elde, R. 1995a. δ-opioid receptor immunoreactivity: distribution in brainstem and spinal cord, and relationship to biogenic amines and enkephalin. *J Neurosci* 15(2):1215–35.

Arvidsson, U., Riedel, M., Chakrabarti, S., Lee, J.-H., Nakano, A., Dado, R., Loh, H., Law, P.-Y., Wessendorf, M.W., and Elde, R. 1995b. Distribution and targeting of a μ-opioid receptor (MOR-1) in brain and spinal cord. *J Neurosci* 15(5):3328–41.

Aston-Jones, G., Ennis, M., Pieribone, V.A., Nickell, W.T., and Shipley, M.T. 1986. The brain nucleus locus coeruleus: restricted afferent control of a broad efferent network. *Science* 234:734–37.

Aston-Jones, G., Rajkowski, J., Kubiak, P., Valentino, R., and Shipley, M. 1996. Role of the locus coeruleus in emotional activation. *Prog Brain Res* 107:379–402.

Atweh, S.F., and Kuhar, M.J. 1977a. Autoradiographic localization of opiate receptors in rat brain. I. Spinal cord and lower medulla. *Brain Res* 124:53–67.

————. 1977b. Autoradiographic localization of opiate receptors in the brain. II. The brainstem. *Brain Res* 129:1–12.

Azami, J., Llewelyn, M.B., and Roberts, M.H.T. 1982. The contribution of nucleus reticularis paragigantocellularis and nucleus raphe magnus to the analgesia produced by systemically administered morphine, investigated with the microinjection technique. *Pain* 12:229–46.

Baamonde, A.I., Hidalgo, A., and Andres-Trelles, F. 1989. Sex-related differences in the effects of morphine and stress on visceral pain. *Neuropharmacology* 28:967–70.

Badillo-Martinez, D., Kirchgessner, A.L., Butler, P.D., and Bodnar, R.J. 1984a. Monosodium glutamate and morphine analgesia: test-specific effects. *Neuropharmacology* 23:1141–49.

Badillo-Martinez, D., Nicotera, N., Butler, P.D., Kirchgessner, A.L., and Bodnar, R.J. 1984b. Impairments in analgesic, hypothermic and glucoprivic stress responses following neonatal monosodium glutamate. *Neuroendocrinology* 38:438–46.

Bajac, D., and Proudfit, H.K. 1999. Projections of neurons in the periaqueductal gray to pontine and medullary catecholamine cell groups involved in the modulation of nociception. *J Comp Neurol* 405:359–79.

Balagura, S., and Ralph, T. 1973. The analgesic effect of electrical stimulation of the diencephalon and mesencephalon. *Brain Res* 60:369–81.

Balthazart, J., Absil, P., Gerard, M., Appeltants, D., and Ball, G. 1998. Appetitive and consummatory male sexual behavior in Japanese quail are differentially regulated by subregions of the preoptic medial nucleus. *J Neurosci* 18:6512–27.

Bandler, R. 1977. Predatory behavior in the cat elicited by lower brainstem and hypothalamic stimulation: a comparison. *Brain Behav Evol* 14:440–60.

Bandler, R., and Flynn, J. 1971. Visual patterned reflex present during hypothalamically elicited attack. *Science* 171:817.

———. 1972. Control of somatosensory fields for striking during hypothalamically elicited attack. *Brain Res* 38:197.

Bandler, R., and Keay, K.A. 1996. Columnar organization in the midbrain periaqueductal gray and the integration of emotional expression. *Prog Brain Res* 107:285–300.

Bandler, R., and Shipley, M.T. 1994. Columnar organization in the midbrain periaqueductal gray: modules for emotional expression. *Trends Neurosci* 17:379–89.

Bandler, R., Price, J., and Keay, K. 2000. Brain mediation of active and passive emotional coping. *Prog Brain Res* 122:333–49.

Banerjee, P., Chatterjee, T., and Ghosh, J.J. 1983. Ovarian steroids and modulation of morphine-induced analgesia and catalepsy in female rats. *Eur J Pharmacol* 96:291–94.

Barbaro, N.M., Heinricher, M.M., and Fields, H.L. 1986. Putative pain modulating neurons in rostral ventral medulla: reflex-related activity predicts effects of morphine. *Brain Res* 366:203–10.

Barden, N., Dupont, A., Labrie, F., Merand, Y., Roulou, D., and Vaudry, H. 1981. Age-dependent changes in the beta-endorphin content of discrete rat brain nuclei. *Brain Res* 208:209–12.

Baron, S.A., and Gintzler, A.R. 1984. Pregnancy-induced analgesia: Effects of adrenalectomy and glucocorticoid replacement. *Brain Res* 321:341–46.

———. 1987. Effects of hypophysectomy and dexamethasone treatment on plasma β-endorphin and pain threshold during pregnancy. *Brain Res* 418:138–45.

Barondez, S. 1993. *Molecules and Mental Illness.* New York: Freeman.

Bartok, R.E., and Craft, R.M. 1997. Sex differences in opioid antinociception. *J Pharmacol Exp Ther* 282:769–78.

Basbaum, A.I., and Fields, H.L. 1984. Endogenous pain control systems: brainstem spinal pathways and endorphin circuitry. *Annu Rev Neurosci* 7:309–38.

Basbaum, A.I., Marley, N.J.E., O'Keefe, J., and Clanton, C.H. 1977. Reversal of morphine and stimulation-produced analgesia by subtotal spinal cord lesions. *Pain* 3:43–56.

Beach, F.A. 1942. Analysis of mechanisms involved in the arousal, maintenance and manifestation of sexual excitement in male animals. *Psychosom Med* 4:173–98.

Beatty, W.W., and Beatty, P.A. 1970. Hormonal determinants of sex differences in avoidance behavior and reactivity to electric shock in the rat. *J Comp Physiol Psychol* 73:446–55.

Beatty, W.W., and Fessler, R.G. 1976. Ontogeny of sex differences in open-field behavior on sensitivity to electric shock in the rat. *Physiol Behav* 16:413–17.

———. 1977. Gonadectomy and sensitivity to electric shock in the rat. *Physiol Behav* 19:1–7.

Becker, J., and Beer, M. 1986. The influence of estrogen on nigrostriatal dopamine activity: behavioral and neurochemical evidence for both pre- and postsynaptic components. *Behav Brain Res* 19:27–33.

Bederson, J., Fields, H., and Barbaro, N. 1990. Hyperalgesia during naloxone-precipitated withdrawal from morphine is associated with increased on-cell activity in the rostral ventromedial medulla. *Somatosens Motor Res* 7:185–203.

Behbehani, M. 1995. Functional characteristics of the midbrain periaqueductal gray. *Prog Neurobiol* 46:575–605.

Behbehani, M.M., and Fields, H.L. 1979. Evidence that an excitatory connection between the periaqueductal gray and the nucleus raphe magnus mediates stimulation produced analgesia. *Brain Res* 170:85–93.

Behbehani, M.M., and Pomeroy, S.L. 1978. Effect of morphine injected into the periaqueductal gray on the activity of single units in nucleus raphe magnus of the rat. *Brain Res* 149:266–69.

Behbehani, M.M., Jiang, M., and Chandler, S.D. 1990. The effect of [Met]enkephalin on the periaqueductal gray neurons of the rat: an in vitro study. *Neuroscience* 38(2):373–80.

Beitz, A.J. 1982a. The nuclei of origin of brainstem enkephalin and substance P projections to the rodent nucleus raphe magnus. *Neuroscience* 7:2753–68.

———. 1982b. The sites of origin of brainstem neurotensin and serotonin projections to the rodent nucleus raphe magnus. *J Neurosci* 2:829–42.

———. 1982c. The organization of afferent projections to the midbrain periaqueductal gray of the rat. *Neuroscience* 7(1):133–59.

———. 1990. Relationship of glutamate and aspartate to the periaqueductal gray–raphe magnus projection: analysis using immunocytochemistry and microdialysis. *J Histochem Cytochem* 38:1755–65.

———. 1995. Periaqueductal gray. In *The Rat Nervous System*, ed. G. Paxinos, 173–82. New York: Academic.

Beitz, A.J., Mullett, M.A., and Weiner, L.L. 1983. The periaqueductal gray projections to the rat spinal trigeminal, raphe magnus, gigantocellualr pars α, and paragigantocellular nuclei arise from separate neurons. *Brain Res* 288:307–14.

Bellgowan, P.S., and Helmstetter, F.J. 1996. Neural systems for the expression of hypoalgesia during nonassociative fear. *Behav Neurosci* 110:727–36.

Ben-Eliyahu, S., Marek, P., Vaccarino, A.L., Mogil, J.S., Sternberg, W.F., and Liebeskind, J.C. 1992. The NMDA receptor antagonist, MK-801, prevents long-lasting non-associative morphine tolerance in the rat. *Brain Res* 575:304–8.

Berkley, K. 1997. Sex differences in pain. *Behav Brain Sci* 20:371–80.

Berkowitz, B.A., and Sherman, S. 1982. Characterization of vasopressin analgesia. *J Pharmacol Exp Ther.* 220:329–34.

Bermant, G. 1961. Response latencies of female rats during sexual intercourse. *Science* 133:1771–73.

Bermant, G., and Westbrook, W.H. 1966. Peripheral factors in the regulation of sexual contact by female rats. *J Comp Physiol Psychol* 61(2):244–50.

Berntson, G.G., and Berson, B.S. 1980. Antinociceptive effects of intraventricular or systemic administration of vasopressin in the rat. *Life Sci* 26:455–59.

Berson, B.S., Berntson, G.G., Zipf, W., Torello, M.W., and Kirk, W.T. 1983. Vasopressin-induced antinociception: an investigation into its physiological and hormonal basis. *Endocrinology* 113:337–43.

Bethea, C. 1993. Colocalization of progestin receptors with serotonin in raphe neurons of macaque. *Neuroendocrinology* 57:1–6.

Beyer, C. 1998. Estrogens stimulate the differentiation of midbrain neurons via a "nongenomic" signalling mechanism. *Eur Neurosci Assoc* Abstr. 60.01:163.

Bhargava, H., Kumar, S., and Bian, J. 1997. Up-regulation of brain N-methyl-D-aspartate receptors following multiple intracerebroventricular injections of [D-Pen2, D-Pen5] enkephalin and [D-Ala2, Glu4] deltorphin II in mice. *Peptides* 18:1609–13.

Bicknell, J. 1999. Estrogens addressing brainstem noradrenergic neurons with projections to hypothalamus. In *Stress and Neuroendocrine Systems*, ed. H. Yamashita. Heidelberg: Springer–Verlag.

Bilsky, E.J., Bernstein, R.N., Pasternak, G.W., Hruby, V.J., Patel, D., Porreca, F., and Lai, J. 1994. Selective inhibition of [D-Ala2,Glu4] deltorphin antinociception by supraspinal, but not spinal, administration of an antisense oligodeoxynucleotide to an opioid δ receptor. *Life Sci* 55:PL37–43.

Bindra, D. 1968. Neuropsychological interpretation of the effects of drive on general activity and instrumental behavior. *Psychol Rev* 75:1–22.

Bindra, D., ed. 1969. *The Interrelated Mechanisms of Reinforcement and Motivation, and the Nature of Their Influence on Response*. Nebraska Symposium on Motivation. Lincoln: University of Nebraska Press.

Blackburn, J.R., Pfaus, J.G., and Phillips, A.G. 1992. Dopamine functions in appetitive and defensive behaviours. *Prog Neurobiol* 39:247–79.

Bland, R., Orn, H., and Newman, S. 1988. Lifetime prevalence of psychiatric disorders in Edmonton. *Acta Psychiatr Scand* 77:24–32.

Bloch, M., Schmidt, P., Danaceau, M., Murphy, J., Nieman, L., and Rubinow, D. 2000. Effects of gonadal steroids in women with a history of postpartum depression. *Am J Psychiatry* 157:924–30.

Bodnar, R.J. 1984. Types of stress that induce analgesia. In *Stress-Induced Analgesia*, ed. M.D. Tricklebank, P.H. Huston, and G. Curzon, 19–32. New York: Wiley.

———. 1986. Neuropharmacological and neuroendocrine substrates of stress-induced analgesia. *Ann NY Acad Sci* 467:345–60.

———. 1996. Opioid receptor subtype antagonists and ingestion. In *Drug Receptor Subtypes and Ingestive Behaviour*, ed. S.J. Cooper and P.J. Clifton, 127–46. London: Academic Press.

Bodnar, R.J., Abrams, G.M., Zimmerman, E.A., Krieger, D.T., Nicholson, G., and Kizer, J.S. 1980a. Neonatal monosodium glutamate: effects upon analgesic responsivity and immunocytochemical ACTH/beta-endorphin. *Neuroendocrinology* 30:280–84.

Bodnar, R.J., and Komisaruk, B.R. 1984. Reduction in cervical probing analgesia by repeated exposure to cold-water swims. *Physiol Behav* 32:653–55.

Bodnar, R.J., and Nicotera, N. 1982. Neuroleptic and analgesic interactions upon pain and activity measures. *Pharmacol Biochem Behav* 16:411–16.

Bodnar, R.J., and Sikorszky, V. 1983. Naloxone and cold-water swim analgesia: parametric considerations and individual differences. *Learn Motivat* 14:223–37.

Bodnar, R.J., Glusman, M., Brutus, M., Spiaggia, A., and Kelly, D.D. 1979a. Analgesia induced by cold-water stress: attenuation following hypophysectomy. *Physiol Behav* 23:53–62.

Bodnar, R.J., Kelly, D.D., and Glusman, M. 1979. 2-deoxy-D-glucose analgesia: influences of opiate and non-opiate factors. *Pharmacol Biochem Behav* 11:297–301.

Bodnar, R.J., Kelly, D.D., Brutus, M., and Glusman, M. 1980b. Stress-induced analgesia: neural and hormonal determinants. *Neurosci Biobehav Rev* 4:87–100.

Bodnar, R.J., Kelly, D.D., Brutus, M., Greenman, C.B., and Glusman, M. 1980c. Reversal of stress-induced analgesia by apomorphine, but not amphetamine. *Pharmacol Biochem Behav* 13:171–75.

Bodnar, R.J., Kelly, D.D., Brutus, M., Mansour, A., and Glusman, M. 1978a. 2-deoxy-D-glucose-induced decrements in operant and reflex pain thresholds. *Pharmacol Biochem Behav* 9:543–49.

Bodnar, R.J., Kelly, D.D., Mansour, A., and Glusman, M. 1979b. Differential effects of hypophysectomy upon analgesia induced by two glucoprivic stressors and morphine. *Pharmacol Biochem Behav* 11:303–8.

Bodnar, R.J., Kelly, D.D., Spiaggia, A., Ehrenberg, C., and Glusman, M. 1978b. Dose-dependent reductions by naloxone of analgesia induced by cold-water swims. *Pharmacol Biochem Behav* 8:667–72.

Bodnar, R.J., Kelly, D.D., Steiner, S.S., and Glusman, M. 1978c. Stress-produced analgesia and morphine-produced analgesia: lack of cross-tolerance. *Pharmacol Biochem Behav* 8:661–66.

Bodnar, R.J., Kordower, J.H., Wallace, M.M., and Tamir, H. 1981. Stress and morphine analgesia: alterations following p-chlorophenylalanine. *Pharmacol Biochem Behav* 14:645–51.

Bodnar, R.J., Lattner, M., and Wallace, M.M. 1980. Antagonism of stress-induced analgesia by D-phenylalanine, an anti-enkephalinase. *Pharmacol Biochem Behav* 13:829–33.

Bodnar, R.J., Mann, P.E., and Stone, E.A. 1985. Potentiation of cold-water swim analgesia by acute, but not chronic, desipramine administration. *Pharmacol Biochem Behav* 23:749–52.

Bodnar, R.J., Merrigan, K.P., and Sperber, E.S. 1983. Potentiation of cold-water swim analgesia and hypothermia by clonidine. *Pharmacol Biochem Behav* 19:447–51.

Bodnar, R.J., Paul, D., and Pasternak, G.W. 1991. Synergistic analgesic interactions between the periaqueductal gray and the locus coeruleus: studies with the partial mu-1 agonist ethylketocyclazocine. *Brain Res* 558:224–30.

Bodnar, R.J., Romero, M.-T., and Kramer, E. 1988. Organismic variables and pain inhibition: roles of gender and aging. *Brain Res Bull* 21:947–53.

Bodnar, R.J., Truesdell, L.S., and Nilaver, G. 1985. Potentiation of vasopressin analgesia in rats treated neonatally with monosodium glutamate. *Peptides* 6:621–26.

Bodnar, R.J., Williams, C.L., Lee, S.J., and Pasternak, G.W. 1988. Role of $\mu 1$ opiate receptors in supraspinal opiate analgesia: a microinjection study. *Brain Res* 447:25–34.

Bodnar, R.J., Zimmerman, E.A., Nilaver, G., Mansour, A., Thomas, L.W., Kelly, D.D., and Glusman, M. 1980d. Dissociation of cold-water swim and morphine analgesia in Brattleboro rats with diabetes insipidus. *Life Sci* 26:1581–90.

Bolles, R.C. 1961. *Theory of Motivation.* New York: Harper & Row.

Bolles, R.C., and Fanselow, M.S. 1980. A perceptual-defensive-recuperative model of fear and pain. *Behav Brain Sci* 3:291–323.

———. 1982. Endorphins and behavior. *Annu Rev Psychol* 33:87–100.

Boutrel, B., Franc, B., Hen, R., Hamon, M., and Adrien, J. 1999. Key role of 5-$HT_{1B}$ receptors in the regulation of paradoxical sleep as evidenced in 5-$HT_{1B}$ knock-out mice. *J Neurosci* 19:3204–12.

Boyer, J.S., Morgan, M.M., and Craft, R.M. 1998. Microinjection of morphine into the rostralventromedial medulla produces greater antinociception in male compared to female rats. *Brain Res* 796:315–18.

Brandling-Bennett, E., Blasberg, M., and Clark, A. 1999. Paced mating behavior in female rats in response to different hormone priming regimens. *Horm Behav* 35:144–54.

Brodie, M.S., and Proudfit, H.K. 1984. Hypoalgesia induced by the local injection of carbachol into the nucleus raphe magnus. *Brain Res* 291:337–42.

———. 1986. Antinociception induced by local injections of carbachol into nucleus raphe magnus: alterations by intrathecal injection of monoaminergic antagonists. *Brain Res* 371:70–79.

Brown, H., Parhar, I., Brooks, P., and Pfaff, D. 1993. Estrogen's induction of preproenkephalin (PPE) mRNA in ventromedial hypothalamus (VMN) of female rats is not augmented by voluntary exercise. *Soc Neurosci Abstr. 19:*485.

Brutus, M., and Siegel, A. 1989. Effects of the opiate antagonist naloxone upon hypothalamically elicited affective defense behavior in the cat. *Behav Brain Res* 33:23–32.

Brutus, M., Shaikh, M., and Siegel, A. 1985. Differential control of hypothalamically elicited flight behavior by the midbrain periaqueductal gray in the cat. *Behav Brain Res* 17:235–44.

Budai, D., and Fields, H.L. 1998. Endogenous opioid peptides acting at $\mu$-opioid receptors in the dorsal horn contribute to midbrain modulation of spinal nociceptive neurons. *J Neurophysiol* 79:677–87.

Budai, D., Harasawa, I., and Fields, H. 1998. Midbrain periaqueductal gray (PAG) inhibits nociceptive inputs to sacral dorsal horn nociceptive neurons through alpha-2-adrenergic receptors. *J Neurophysiol* 80:2244–54.

Bueno, J., and Pfaff, D. 1976. Single unit recording in hypothalamus and preoptic area of estrogen-treated and untreated ovariectomized female rats. *Brain Res* 101:67–78.

Bunzow, J., Saez, C., Mortrud, M., Bouvier, C., Williams, J., Low, M., and Grandy, D. 1994. Molecular cloning and tissue distribution of a putative member of the rat opioid receptor gene family that is not a $\mu$, $\delta$ or $\kappa$ receptor type. *FEBS Lett* 347:284–88.

Burleson, M., Malarkey, W., Cacioppo, J., Poehlmann, K., Kiecolt-Glaser, J., Berntson, G., and Glaser, R. 1998. Postmenopausal hormone replacement: effects on autonomic, neuroendocrine, and immune reactivity to brief psychological stressors. *Psychosom Med* 60:17–25.

Bush, F., Harkins, S., Harrinton, W., and Price, D. 1993. Analysis of gender effects on pain perception and symptom presentation in temporomandibular pain. *Pain* 53:73–80.

Butler, P.D., and Bodnar, R.J. 1984. Potentiation of footshock analgesia by thyrotropin releasing hormone. *Peptides* 5:635–39.

———. 1987. Neuromodulatory effects of TRH upon swim and cholinergic analgesia. *Peptides* 8:299–307.

Caggiula, A., Shaw, D., Antelman, S., and Edwards, D. 1976. Interactive effect of brain catecholamines and variations in sexual and nonsexual arousal on copulatory behavior of male rats. *Brain Res* 111:321–26.

Cameron, A.A., Khan, I.A., Westlund, K.N., and Willis, W.D. 1995. The efferent projections of the periaqueductal gray of the rat: A Phaseolus vulgaris-leucoagglutinin study. II. Descending projections. *J Comp Neurol* 351:585–601.

Cameron, D.L., Wessendorf, M.W., and Williams, J.T. 1997. A subset of ventral tegmental area neurons is inhibited by dopamine, 5-hydroxytryptamine and opioids. *Neuroscience* 77(1):155–66.

Campbell, S., and Whitehead, M. 1977. Oestrogen therapy and the menopausal syndrome. *Clin Obstet Gynaecol* 4:31–47.

Candido, J., Lufty, K., Billings, B., Sierra, V., Duttaroy, A., and Inturrisi, C.E. 1992. Effect of adrenal and sex hormones on opioid analgesia and opioid receptor regulation. *Pharmacol Biochem Behav* 42:685–92.

Cannon, J., Prieto, G., Lee, A., and Liebeskind, J. 1982. Evidence for opioid and nonopioid forms of stimulation-produced analgesia in the rat. *Brain Res* 243:315–21.

Cannon, J.T., Lewis, J.W., Weinberg, V.E., and Liebeskind, J.C. 1983. Evidence for the independence of brainstem mechanisms mediating analgesia induced by morphine and two forms of stress. *Brain Res* 269:231–36.

Canteras, N.S., Simerly, R.B., and Swanson, L.W. 1994. Organization of projections from the ventromedial nucleus of the hypothalamus: a Phaseolus vulgaris-leucoagglutinin study in the rat. *J Comp Neurol* 348(1):41–79.

Cape, E., and Jones, B. 1998. Differential modulation of high-frequency γ-electroencephalogram activity and sleep-wake state by noradrenaline and serotonin microinjections into the region of cholinergic basalis neurons. *J Neurosci* 18:2653–66.

Catelli, J.M., Sved, A.F., and Komisaruk, B.R. 1987. Vaginocervical probing elevates blood pressure and induces analgesia by separate mechanisms. *Physiol Behav* 41:609–12.

Caudle, R.M., and Isaac, L. 1988. Influence of dynorphin(1–13) on spinal reflexes in the rat. *J Pharmacol Exp Ther* 246:508–13.

Cedarbaum, J., and Aghajanian, G. 1978. Activation of locus coeruleus neurons by peripheral stimuli: modulation by a collateral inhibitory mechanism. *Life Sci* 23:1383–92.

Chakrabarti, S., Rivera, M., Yan, S., Tang, W., and Gintzler, A. 1998a. Chronic morphine augments $G(\beta)(\gamma)/Gs(\alpha)$ stimulation of adenylyl cyclase: relevance to opioid tolerance. *Mol Pharmacol* 54:655–62.

Chakrabarti, S., Wang, L., Tang, W., and Gintzler, A. 1998b. Chronic morphine augments adenylyl cyclase phosphorylation: relevance to altered signaling during tolerance/dependence. *Mol Pharmacol* 54:949–53.

Chan, S.H., and Lai, Y.Y. 1982. Effects of aging on pain responses and analgesic efficacy of morphine and clonidine in rats. *Exp Neurol* 75:112–19.

Chance, W.T. 1980. Autoanalgesia: opiate and non-opiate mechanisms. *Neurosci Biobehav Rev* 4:55–67.

Chance, W.T., White, A.C., Krynock, G.M., and Rosecrans, J.A. 1977. Autoanalgesia: behaviorally activated antinociception. *Eur J Pharmacol* 44:283–84.

———. 1978. Conditional fear-induced antinociception and decreased binding of ($^3$H) leu-enkephalin to rat brain. *Brain Res* 141:371–74.

Chang, P., Aicher, S., and Drake, C. 2000. κ opioid receptors in rat spinal cord vary across the estrous cycle. *Brain Res* 861:168–72.

Chatterjee, T.K., Das, S., Banerjee, P., and Ghosh, J.J. 1982. Possible physiological role of adrenal and gonadal steroids in morphine analgesia. *Eur J Pharmacol* 77:119–21.

Chavkin, C., James, I.F., and Goldstein, A. 1982. Dynorphin is a specific endogenous ligand of the κ receptor. *Science* 215:413–15.

Chen, L., and Huang, L.Y. 1991. Sustained potentiation of NMDA receptor-mediated glutamate responses through activation of protein kinase C by a μ opioid. *Neuron* 7(2):319–26.

———. 1992. Protein kinase C reduces $Mg^{2+}$ block of NMDA-receptor channels as a mechanism of modulation. *Nature* 356(6369):521–23.

Chen, X.H., Adams, J.U., Geller, E.B., DeRiel, J.K., Adler, M.W., and Liu-Chen, L.Y. 1995. An antisense oligodeoxynucleotide to μ-opioid receptors inhibits μ-opioid receptor agonist-induced analgesia in rats. *Eur J Pharmacol* 275: 105–8.

Cheng, Z., Fields, H., and Heinricher, M. 1986. Morphine microinjected into the periaqueductal gray has differential effects on three classes of medullary neurons. *Brain Res* 375:57–65.

Chien, C.C., Brown, G.P., Pan, Y.X., and Pasternak, G.W. 1994. Blockade of U50488H analgesia by antisense oligodeoxynucleotides to a κ opioid receptor. *Eur J Pharmacol* 253:R7–8.

Cheng, B., and Christie, M.J. 1996. Local opioid withdrawal in rat single periaqueductal gray neurons in vitro. *J. Neurosci* 16:7128–36.

Cho, H., and Basbaum, A. 1989. Ultrastructural analysis of dynorphin B-immunoreactive cells and terminals in the superficial dorsal horn of the deafferented spinal cord of the rat. *J Comp Neurol* 281:193–205.

Choleris, E., and Kavaliers, M. 1999. Social learning in animals: sex differences and neurobiological analysis. *Pharmacol Biochem Behav* 64:767–76.

Christensen, B., and Perl, E. 1970. Spinal neurons specifically excited by noxious or thermal stimuli: marginal zone of the dorsal horn. *J Neurophysiol* 33: 293–307.

Christie, M.J., Chesher, G.B., and Bird, K.D. 1981. The correlation between swim-stress induced antinociception and ($^3$H) leu-enkephalin binding to brain homogenates in mice. *Pharmacol Biochem Behav* 15:853–57.

Christie, M.J., Trisdikoon, P., and Chesher, G.B. 1982. Tolerance and cross-tolerance with morphine resulting from physiological release of endogenous opiates. *Life Sci* 31:839–45.

Chu, T.G., and Orlowski, M. 1985. Soluble metalloendopeptidase from rat brain: action on enkephalin-containing peptides and other bioactive peptides. *Endocrinology* 116:1418–25.

Cicero, T.J., Nock, B., and Meyer, E.R. 1996. Gender-related differences in the antinociceptive properties of morphine. *J Pharmacol Exp Ther* 279:767–73.

———. 1997. Sex-related differences in morphine's antinociceptive activity: relationship to serum and brain morphine concentrations. *J Pharmacol Exp Ther* 282:939–44.

Clark, F.M., and Proudfit, H.K. 1991a. The projection of locus coeruleus neurons to the spinal cord in the rat determined by anterograde tracing combined with immunocytochemistry. *Brain Res* 538:231–45.

———. 1991b. The projection of noradrenergic neurons in the A7 cate-cholamine cell group to the spinal cord in the rat demonstrated by antero-grade tracing combined with immunocytochemistry. *Brain Res* 547:279–88.

———. 1991c. Projections of neurons in the ventromedial medulla to pontine catecholamine cell groups involved in the modulation of nociception. *Brain Res* 540:105–15.

Clarke, I., Scott, C., Pereira, A., and Rawson, J. 1999. Levels of dopamine β hy-droxylase immunoreactivity in the preoptic hypothalamus of the ovariectom-ised ewe following injection of oestrogen: evidence for increased noradren-aline release around the time of the oestrogen-induced surge in luteinizing hormone. *J Neuroendocrinol* 11:503–12.

Clements, J.R., Madl, J.E., Johnson, R.L., Larson, A.A., and Beitz, A.J. 1987. Lo-calization of glutamate, glutaminase, aspartate and aspartate aminotrans-ferase in the rat midbrain periaqueductal gray. *Exp Brain Res* 67:594–602.

Cogan, R., and Spinnato, J.A. 1986. Pain and discomfort thresholds in late preg-nancy. *Pain* 27:63–68.

Cohen, G.A., Doze, V.A., and Madison, D.V. 1992. Opioid inhibition of GABA release from presynaptic terminals of rat hippocampal interneurons. *Neu-ron* 9(2):325–35.

Comb, M., Herbert, E., and Crea, R. 1982. Partial characterization of the mRNA that codes for enkephalins in bovine adrenal medulla and human pheo-chromocytoma. *Proc Natl Acad Sci USA* 79:360–64.

Commons, K., and Milner, T. 1997. Localization of delta opioid receptor im-munoreactivity in interneurons and pyramidal cells in the rat hippocampus. *J Comp Neurol* 381(3):373–87.

Commons, K., and Pfaff, D. 2001. Anatomical basis for enkephalin mediated dis-inhibition in the ventromedial nucleus of the hypothalamus. *Neuroendocri-nology*. In press.

———. 2001. Ultrastructural evidence for enkephalin mediated disinhibition in the ventromedial nucleus of the hypothalamus. *J Chem Neuroanat* 21:53–62.

Commons, K.G., Aicher, S.A., Kow, L.-M., and Pfaff, D.W. 2000. Presynaptic and postsynaptic distribution of the mu opioid receptor to GABAergic and medullary projecting periaqueductal gray neurons. *J Comp Neurol*. In press.

Commons, K.G., and Milner, T.A. 1996. Ultrastructural relationships between leu-enkephalin-and GABA-containing neurons differ within the hippocam-pal formation. *Brain Res* 724(1):1–15.

Commons, K., Kow, L., Milner, T., and Pfaff, D. 1999. In the ventromedial nu-cleus of the rat hypothalamus, GABA-immunolabeled neurons are abun-dant and are innervated by both enkephalin-and GABA-immunolabeled axon terminals. *Brain Res* 816:58–67.

Commons, K., Van Bockstaele, E., and Pfaff, D. 1999. Frequent colocalization of mu opioid and NMDA-type glutamate receptors at postsynaptic sites in pe-riacqueductal gray neurons. *J Comp Neurol* 408:549–59.

Conrad, L., and Pfaff, D. 1976a. Efferents from medial basal forebrain and hypothalamus in the rat. I. An autoradiography study of the medial preoptic area. *J Comp Neurol* 169:185–220.

———. 1976b. Efferents from medial basal forebrain and hypothalamus in the rat. II. An autoradiography study of the anterior hypothalamus. *J Comp Neurol* 169:221–62.

Conrad, L., Leonard, C., and Pfaff, D. 1974. Connections of the median and dorsal raphe nuclei in the rat: an autoradiography and degeneration study. *J. Comp. Neurol.* 156:179–206.

Corbett, A.D., Patterson, S.J., McKnight, A.T., Magnan, J., and Kosterlitz, H.W. 1982. Dynorphin (1–8) and dynorphin (1–9) are ligands for the kappa subtype of opiate receptor. *Nature* 299:79–81.

Cortes, R., and Palacios, J.M. 1986. Muscarinic cholinergic receptor subtypes in the rat brain. I. Quantitative autoradiographic studies. *Brain Res* 362:227–38.

Cottingham, S., and Pfaff, D. 1986. Interconnectedness of steroid hormone binding neurons: existence and implications. In *Current Topics of Neuroendocrinology*, vol. 7, ed. D. Ganten and D. Pfaff, 223–50, Heidelberg: Springer Verlag.

Coull, J. 1998. Neural correlates of attention and arousal: insights from electrophysiology, functional neuroimaging and psychopharmacology. *Prog Neurobiol* 55:343–61.

Crabbe, J., Wahlsten, D., and Dudek, B. 1999. Genetics of mouse behavior: interactions with laboratory environment. *Science* 284:1670–72.

Craft, R., and Milholland, R. 1998. Sex differences in cocaine-and nicotine-induced antinociception in the rat. *Brain Res* 809:137–40.

Crow, T., Deakin, J., File, S., Longden, A., and Wendlandt, S. 1978. The locus coeruleus noradrenergic system: evidence against a role in attention, habituation, anxiety and motor activity. *Brain Res* 155:249–61.

Cstonas, K., Rust, M., Hollt, V., Mahr, W., Kromer, W., and Teschmacher, H.J. 1979. Elevated plasma beta-endorphin levels in pregnant women and their neonates. *Life Sci* 25:835–44.

Cunningham, E.T.J., Bohn, M.C., and Sawchenko, P.E. 1990. Organization of adrenergic inputs to the paraventricular and supraoptic nuclei of the hypothalamus in the rat. *J Comp Neurol* 292:651–67.

Curran, M., and Petersen, S. 1998. Regional colocalization of progestin receptor (PR) mRNA with tyrosine hydroxylase (TH) mRNA in brain stem. *Soc Neurosci Abstr* 24 (abstr. 736.4):1853.

Czlonowski, A., Millan, M.J., and Herz, A. 1987. The selective kappa-opioid agonist, U50488H, produces antinociception in the rat via a supraspinal action. *Eur J Pharmacol* 142:183–84.

Dado, R.J., Law, P.Y., Loh, H.H., and Elde, R. 1993. Immunofluorescent identification of a delta-opioid receptor on primary afferent nerve terminals. *Neuroreport* 5:341–44.

Dahlstrom, A., and Fuxe, K. 1964. Evidence for the existence of monamine neurons in the central nervous system. I. Demonstration of monoamines in the cell bodies of brain stem neurons. *Acta Physiol Scand* 62(Suppl. 232):1–55.

Daniels, D., Miselis, R.R., and Flanagan-Cato, L.M. 1999. Central neuronal circuit innervating the lordosis-producing muscles defined by transneuronal transport of pseudorabies virus. *J Neurosci* 19(7):2823–33.

Darkow, D., Lu, L., and White, R. 1997. Estrogen relaxation of coronary artery smooth muscle is mediated by nitric oxide and cGMP. *Am J Physiol* 272: H2765–73.

Davis, P.G., McEwen, B., and Pfaff, D.W. 1979. Localized behavioral effects of tritiated estradiol implants in the ventromedial hypothalamus of female rats. *Endocrinology* 104:893–903.

Davis, P., Zhang, S., and Bandler, R. 1996. Midbrain and medullary regulation of respiration and vocalization. *Prog Brain Res* 107:315–25.

Dawson-Basoa, M., and Gintzler, A. 1997. Involvement of spinal cord delta opiate receptors in the antinociception of gestation and its hormonal stimulation. *Brain Res* 757:37–42.

———. 1998. Gestational and ovarian sex steroid antinociception: synergy between spinal kappa and delta opioid systems. *Brain Res* 794:61–67.

Dawson-Basoa, M.B., and Gintzler, A.R. 1993. 17-beta estradiol and progesterone modulate an intrinsic opioid analgesic system. *Brain Res* 601:241–45.

———. 1996. Estrogen and progesterone activate spinal kappa-opiate receptor analgesic mechanisms. *Pain* 64:607–15.

Dellovade, T., Chan, J.B.V., Forrest, D., and Pfaff, D. 2000. The two thyroid hormone receptor genes have opposite effects on estrogen stimulated sex behaviors. *Nat Neurosci* 3:472–75.

Dellovade, T.L., Zhu, Y.S., and Pfaff, D.W. 1999. Thyroid hormones and estrogen affect oxytocin gene expression in hypothalamic neurons. *J Neuroendocrinol* 11:1–10.

Depaulis, A., Morgan, M.M., and Liebeskind, J.C. 1987. GABAergic modulation of the analgesic effects of morphine microinjected in the ventral periaqueductal gray matter of the rat. *Brain Res* 436(2):223–28.

Dostrovsky, J.O., and Deakin, J.F.W. 1977. Periaqueductal gray lesions reduce morphine analgesia in the rat. *Neurosci Lett* 4:99–103.

Dourish, C., and Iversen, S. 1989. Blockade of apomorphine induced yawning in rats by the dopamine autoreceptor antagonist (+)-AJ 76. *Neuropharmacology* 28:1423–25.

Drake, C.T., and Milner, T.A. 1999. Mu opioid receptors are in somatodendritic and axonal compartments of GABAergic neurons in rat hippocampal formation. *Brain Res* 849(1–2):203–15.

Dringenberg, H., and Vanderwolf, C. 1996. 5-hydroxytryptamine (5-HT) agonists: effects on neocortical slow wave activity after combined muscarinic and serotonergic blockade. *Brain Res* 728:181–87.

————. 1997. Neocortical activation: modulation by multiple pathways acting on central cholinergic and serotonergic systems. *Exp Brain Res* 116:160–74.

————. 1998. Involvement of direct and indirect pathways in electrocortico-graphic activation. *Neurosci Biobehav Rev* 22:243–57.

Drugan, R.C., Grau, J.W., Maier, S.F., Madden, J., and Barchas, J.D. 1981. Cross tolerance between morphine and the long-term analgesic reaction to inescapable shock. *Pharmacol Biochem Behav* 14:677–82.

Drury, R.A., and Gold, R.M. 1978. Differential effects of ovarian hormones on reactivity to electric footshock in the rat. *Physiol Behav* 20:187–91.

Dudley, C., and Moss, R. 1985. LHRH and mating behavior: sexual receptivity versus sexual preference. *Pharmacol Biochem Behav* 22:967–72.

Duggan, A., and Morton, C. 1988. Tonic descending inhibition and spinal nociceptive transmission. *Prog Brain Res* 77:193–207.

Duggan, A.W., and North, R.A. 1983. Electrophysiology of opioids. *Pharmacol Rev* 35(4):219–81.

Dupont, A., Savard, P., Merand, Y., Labrie, F., and Rossier, J. 1981. Age-related changes in central nervous system enkephalin and substance P. *Life Sci* 29:2317–2322.

Eckersell, C.B., Popper, P., and Micevych, P.E. 1998. Estrogen-induced alteration of μ-opioid receptor immunoreactivity in the medial preoptic nucleus and medial amygdala. *J Neurosci* 18:3967–76.

Edwards, H., Burnham, W., and MacLusky, N. 1999. Testosterone and its metabolites affect afterdischarge thresholds and the development of amygdala kindled seizures. *Brain Res* 838:151–57.

Edwards, H., Burnham, W., Mendonca, A., Bowlby, D., and MacLusky, N. 1999a. Steroid hormones affect limbic afterdischarge thresholds and kindling rates in adult female rats. *Brain Res* 838:136–50.

Edwards, H., Burnham, W., Ng, M., Asa, S., and MacLusky, N. 1999b. Limbic seizures alter reproductive function in the female rat. *Epilepsia* 40:1370–77.

Egger, M., and Flynn, J. 1962. Amygdaloid suppression of hypothalamically elicited attack behavior. *Science* 136:43–44.

————. 1963. Effects of electrical stimulation of the amygdala on hypothalamically elicited attack behavior in cats. *J Neurophysiol* 26:705.

Ellermeier, W., and Westphal, W. 1994. Gender differences in pain ratings and pupil reactions to painful pressure stimuli. *Pain* 61:435–39.

Ellis, L., and Ebertz, L. 1998. *Males, Females, and Behavior: Toward Biological Understanding.* Westport, Conn.: Praeger.

Emery, D., and Moss, R. 1984. Lesions confined to the ventromedial hypothalamus decrease the frequency of coital contacts in female rats. *Horm Behav* 18:313–29.

Ennis, M., Shipley, M.T., Behbehani, M., VanBockstaele, E.J., and Aston-Jones, G. 1991. Projections from the periaqueductal gray to nucleus locus coeruleus and pericoerulear region: anatomic and physiologic studies. *J Comp Neurol* 306:480–94.

Ernst, C., and Angst, J. 1992. Sex differences in depression: evidence from longitudinal epidemiological data. *Eur Arch Psychiatry Clin Neurosci* 241:222–30.

Erskine, M.S. 1989. Solicitation behavior in the estrous female rat: a review. *Horm Behav* 23:373–502.

Evans, C., Keith, D., Morrison, H., Magendzo, K., and Edwards, R. 1992. Cloning of a δ-opioid receptor by functional expression. *Science* 258:1952–55.

Eysenck, M. 1982. *Attention and Arousal.* Berlin: Springer-Verlag.

Facchinetti, F., Centin, G., Parina, D., Petraglia, F., D'Antona, N., Cosmi, E.V., and Genazzani, A.R. 1982. Opioid plasma levels during labour. *Gynecol Obstet Invest* 13:155–63.

Faden, A.I., and Jacobs, T.P. 1983. Dynorphin produces partially reversible paraplegia in the rat. *Eur J Pharmacol* 91:321–24.

Fahrbach, S.E., Morrell, J.I., and Pfaff, D.W. 1989. Studies of ventromedial hypothalamic afferents in the rat using three methods of HRP application. *Exp Brain Res* 77(2):221–33.

Fahrbach, S., Meisel, R., and Pfaff, D. 1985. Preoptic implants of estradiol increase wheel running but not the open field activity of female rats. *Physiol Behav* 35:985–92.

Fahrbach, S., Morrell, J.I., and Pfaff, D.W. 1986. Identification of medial preoptic neurons that concentrate estradiol and project to the midbrain in the rat. *J Comp Neurol* 247:364–82.

Fang, F., and Proudfit, H. 1996. Spinal cholinergic and monoamine receptors mediate the antinociceptive effect of morphine in the periaqueductal gray on the rat tail, but not the feet. *Brain Res* 722:95–108.

———. 1998. Antinociception produced by microinjection of morphine in the rat periaqueductal gray is enhanced in the foot, but not the tail, by intrathecal injection of α-1-adrenoceptor antagonists. *Brain Res* 790:14–24.

Fang, F.G., Fields, H.L., and Lee, N.M. 1986. Action at the mu receptor is sufficient to explain supraspinal analgesic effect of opiates. *J Pharmacol Exp Ther* 238:1039–44.

Fanselow, M. 1986. Conditioned fear-induced opiate analgesia: a competing motivational state theory of stress induced analgesia. *Ann NY Acad Sci* 467:40–54.

Fanselow, M., and Sigmundi, R. 1987. Functional behaviorism and aversively motivated behavior: a role for endogenous opioids in the defensive behavior of the rat. *Psychol Rec* 37:317–34.

Fardin, V., Oliveras, J.L., and Besson, J.M. 1984a. Projections from the periaqueductal gray matter to the B-3 cellular area (nucleus raphe magnus and nucleus reticularis paragigantocellularis) as revealed by the retrograde transport of horseradish peroxidase in the rat. *J Comp Neurol* 223: 483–500.

———. 1984b. A reinvestigation of the analgesic effects induced by stimulation of the periaqueductal gray matter in the rat. I. The production of behavioral side effects together with analgesia. *Brain Res* 306:105–23.

———. 1984c. A reinvestigation of the analgesic effects induced by stimulation of the periaqueductal gray matter in the rat. II. Differential characteristics of the analgesia induced by ventral and dorsal PAG stimulation. *Brain Res* 306:125–39.

Faris, P., Komisaruk, B.K., Watkins, L.R., and Mayer, D.J. 1983. Evidence for the neuropeptide cholecystokinin as an antagonist of opiate analgesia. *Science* 219:310–12.

Feine, J., Bushnell, M., Miron, D., and Duncan, G. 1991. Sex differences in the perception of noxious heat stimuli. *Pain* 44:255–62.

Fekete, C., Strutton, P., Cagampang, F., et al. 1999. Estrogen receptor immunoreactivity is present in the majority of central histaminergic neurons: evidence for a new neuroendocrine pathway associated with luteinizing hormone-releasing hormone-synthesizing neurons in rats and humans. *Endocrinology* 140:4335–41.

Fendt, M., and Fanselow, M. 2000. The neuroanatomical and neurochemical basis of conditioned fear. *Neurosci Biobehav Rev* 23:743–60.

Fiber, J., and Etgen, A. 1998. GABA-augmented $^3$H-norepinephrine release is modulated by opioids in brain slices from female rats. *Soc Neurosci Abstr* 24: 2065.

Fiber, J.M., and Etgen, A.M. 1997. GABA augments basal and electrically stimulated 3H–norepinephrine release in hypothalamic, preoptic area and cortical slices of female rats. *Neurochem Int* 31:769–80.

Fields, H. 1992. Is there a facilitating component to central pain modulation? *APSJ* 1:71–78.

Fields, H. 2000. Pain modulation: expectation, opioid analgesia and virtual pain. *Prog Brain Res* 122:245–53.

Fields, H., and Basbaum, A.I. 1978. Brain control of spinal pain transmission neurons. *Annu Rev Physiol* 40:217–48.

Fields, H.L., and Anderson, S.D. 1978. Evidence that raphe-spinal neurons mediate opiate and midbrain stimulation-produced analgesias. *Pain* 5:333–49.

Fields, H.L., Basbaum, A.I., Clanton, C.H., and Anderson, S.D. 1977. Nucleus raphe magnus inhibition of spinal cord dorsal horn neurons. *Brain Res* 126: 441–53.

Fields, H.L., Heinricher, M.M., and Mason, P. 1991. Neurotransmitters in nociceptive modulatory circuits. *Annu Rev Neurosci* 14:219–45.

Fields, H.L., Vanegas, H., Hentall, I., and Zorman, G. 1983. Evidence that disinhibition of brainstem neurons contributes to morphine analgesia. *Nature* 306:684–86.

Fillingim, R., and Maixner, W. 1995. Gender differences in response to noxious stimuli. *Pain Forum* 4:209–21.

Fillingim, R., Edwards, R., and Powell, T. 1999. The relationship of sex and clinical pain to experimental pain responses. *Pain* 83:419–25.

———. 2000. Sex-dependent effects of reported familial pain history on recent pain complaints and experimental pain responses. *Pain* 86:87–94.

Fillingim, R., Maddux, V., and Shackelford, J. 1999. Sex differences in heat pain thresholds as a function of assessment method and rate of rise. *Somatosens Motor Res* 16:57–62.

Fillingim, R., Maixner, W., Girdler, S., Light, K., Harris, M., Sheps, D., and Mason, G. 1997. Ischemic but not thermal pain sensitivity varies across the menstrual cycle. *Psychosom Med* 59:512–20.

Fillingim, R., Maixner, W., Kincaid, S., and Silva, S. 1998. Sex differences in temporal summation but not sensory-discriminative processing of thermal pain. *Pain* 75:121–27.

Fink, G., Dow, R., McQueen, J., Bennie, J., and Carroll, S. 1999. Serotonergic 5-$HT_{2A}$ receptors important for the oestradiol-induced surge of luteinising hormone-releasing hormone in the rat. *J Neuroendocrinology* 11:63–69.

Fink, G., Sumner, B., McQueen, J., Wilson, H., and Rosie, R. 1998. Sex steroid control of mood, mental state and memory. *Clin Exp Pharmacol Physiol* 25:764–75.

Fisher, M.C., and Bodnar, R.J. 1992. 2-deoxy-D-glucose antinociception and serotonin receptor subtype antagonists: test-specific effects. *Pharmacol Biochem Behav* 43:1241–46.

Floody, O., and Pfaff, D. 1974. Steroid hormones and aggressive behavior: approaches to the study of hormone-sensitive brain mechanisms for behavior. *Res Publ, Assn Res Ner Ment Dis* 52:149–85.

———. 1977a. Aggressive behavior in female hamsters: the hormonal basis for fluctuations in female aggressiveness correlated with estrous state. *J Comp Physiol Psychol* 91:443–64.

———. 1977b. Communication among hamsters by high-frequency acoustic signals: I. Physical characteristics of hamster calls. *J Comp Physiol Psychol* 91:794–806.

Floody, O., Pfaff, D., and Lewis, C. 1977. Communication among hamsters by high-frequency acoustic signals: II. Determinants of calling by females and males. *J Comp Physiol Psychol* 91:807–19.

Floyd, N., Keay, K., Arias, C., Sawchenko, P., and Bandler, R. 1996. Projections from the ventrolateral periaqueductal gray to endocrine regulatory subdivisions of the paraventricular nucleus of the hypothalamus in the rat. *Neurosci Lett* 220:105–8.

Flugge, G., Pfender, D., Rudolph, S., Jarry, H., and Fuchs, E. 1999. 5HT$_{1A}$-receptor binding in the brain of cyclic and ovariectomized female rats. *J Neuroendocrinol* 11:243–49.

Flynn, J., Edwards, S., and Bandler, R. 1971. Changes in sensory and motor systems during centrally elicited attack. *Behav Sci* 16:1–19.

Flynn, J., Vanegas, H., Foote, W., and Edwards, J. 1970. Neural mechanisms involved in a cat's attack on a rat. In *The neural control of behavior*. ed. P. Whalen, 135–73. New York: Academic.

Fort, P., Khateb, A., Pegna, A., Muhlethaler, M., and Jones, B. 1995. Noradrenergic modulation of cholinergic nucleus basalis neurons demonstrated by in vitro pharmacological and immunohistochemical evidence in the guinea-pig brain. *Eur J Neurosci* 7:1502–11.

Fox, R.J., and Sorenson, C.A. 1994. Bilateral lesions of the amygdala attenuate analgesia induced by diverse environmental challenges. *Brain Res* 648:215–21.

Freeman, E., Purdy, R., Coutifaris, C., Rickels, K., and Paul, S. 1993. Anxiolytic metabolites of progesterone: correlation with mood and performance measures following oral progesterone administration to healthy female volunteers. *Neuroendocrinology* 58:478–84.

French, J.D. 1952. Effects of chronic lesions in central cephalic brain stem of monkeys. *Arch Neurol Psychiatry* 68:591–604.

Friedman, H.J., Jen, M.F., Chang, J.K., Lee, N.M., and Loh, H.H. 1981. Dynorphin: a possible modulatory peptide on morphine or β-endorphin analgesia in the mouse. *Eur J Pharmacol* 69:357–60.

Frohlich, J., Morgan, M., Chan, J., and Pfaff, D. 2001. Hormonal influences on arousal measures in female mice. I. Factor and cluster analyses of arousal components. *Horm Behav*. In press.

Frohlich, J., Ogawa, S., Morgan, M., Burton, L., and Pfaff, D. 1999. Hormones, genes and the structure of sexual arousal. *Behav Brain Res* 105:5–27.

Frye, C.A., Bock, B.C., and Kanarek, R.B. 1992. Hormonal milieu affects tail-flick latency in female rats and may be attenuated by access to sucrose. *Physiol Behav* 52:699–706.

Frye, C., and DeBold, J. 1993. 3a-OH-DHP and 5a-THDOC implants to the VTA facilitate sexual receptivity in hamsters after progesterone priming to the VMH. *Brain Res* 612:130–37.

Frye, C., and Gardiner, S. 1996. Progestins can have a membrane-mediated action in rat midbrain for facilitation of sexual receptivity. *Horm Behav* 30:682–91.

Frye, C., and Leadbetter, E. 1994. 5a-reduced progesterone metabolites are essential in hamster VTA for sexual receptivity. *Life Sci* 54:653–59.

Frye, C.A., Van Keuren, K.R., Rao, P.N., and Erskine, M.S. 1996. Analgesic effects of the neurosteroid 3 α-androstanediol. *Brain Res* 709:1–9.

Futo, J., Shay, J., Block, S., Holt, J., Beach, M., and Moss, J. 1992. Estrogen and progesterone withdrawal increases cerebral vasoreactivity to serotonin in rabbit basilar artery. *Life Sci* 50:1165–72.

Fuxe, K. 1965. Evidence for the existence of monoamine containing neurons in the central nervous system. IV. The distribution of monoamine terminals in the central nervous system. *Acta Physiol Scand* 62(Suppl. 247):39–85.

Fuxe, K., Hokfelt, T., and Ungerstedt, U. 1970. Morphological and functional aspects of central monamine neurons. *Int Rev Neurobiol* 13:93–126.

Gainetdinov, R., Wetsel, W., Jones, S., Levin, E., Jaber, M., and Caron, M. 1999. Role of serotonin in the paradoxical calming effect of psychostimulants on hyperactivity. *Science* 283:397–401.

Gallistel, C.R. 1980. *The Organization of Action: A New Synthesis*. Hilldale, N.J.: Erlbaum.

Gambert, S.R. 1981. Interaction of age and thyroid hormone status on β-endorphin content in rat corpus striatum and hypothalamus. *Neuroendocrinology* 32:114–17.

Gambert, S.R., Garthwaite, T.L., Pontzer, C.H., and Hagen, T.C. 1980. Age-related changes in central nervous system β-endorphin and ACTH. *Neuroendocrinology* 31:252–55.

Gammie, S., and Truman, J. 1997. Neuropeptide hierarchies and the activation of sequential motor behaviors in the hawkmoth, Manduca sexta. *J Neurosci* 17:4389–97.

Gao, K., Chen, D.O., Genzen, J.R., and Mason, P. 1998. Activation of serotonergic neurons in the raphe magnus is not necessary for morphine analgesia. *J Neurosci* 18:1860–68.

Gao, K., Kim, Y.H., and Mason, P. 1997. Serotonergic pontomedullary neurons are not activated by antinociceptive stimulation in the periaqueductal gray. *J Neurosci* 17:3285–92.

Gear, R., Gordon, N., Heller, P., Paul, S., Miaskowski, C., and Levine, J. 1996a. Gender difference in analgesic response to the κ-opioid pentazocine. *Neurosci Lett* 205:207–9.

Gear, R., Miaskowski, C., Gordon, N., Paul, S., Heller, P., and Levine, J. 1996b. κ-opioids produced significantly greater analgesia in women than in men. *Nat Med* 2:1248–50.

———. 1999. The κ opioid nalbuphine produces gender- and dose-dependent analgesia and antianalgesia in patients with postoperative pain. *Pain* 83: 339–45.

Gebhart, G.F. 1986. Modulatory effects of descending systems on spinal dorsal horn neurons. In *Spinal Afferent Processing*, ed. T.L. Yaksh, 391–416. New York: Plenum.

Gebhart, G., Sandkuhler, J., Thalhammer, J., and Zimmermann, M. 1983a. Quantitative comparison of inhibition in spinal cord of nociceptive information by stimulation in periaqueductal gray or nucleus raphe magnus of the cat. *J Neurophysiol* 50:1433–45.

Gebhart, G.F., Sandkuhler, J., Thalhammer, J.G., and Zimmerman, M. 1983b. Inhibition of spinal nociceptive information by stimulation in midbrain of the cat is blocked by lidocaine microinjected in nucleus raphe magnus and medullary reticular formation. *J Neurophysiol* 50:1446–58.

Gintzler, A., Adapa, I., Toll, L., Medina, V., and Wang, L. 1997. Modulation of enkephalin release by nociceptin (orphanin FQ). *Eur J Pharmacol* 325:29–34.

Gintzler, A., Chan, W., and Glass, J. 1987. Evoked release of methionine enkephalin from tolerant/dependent enteric ganglia: paradoxical dependence on morphine. *Proc Natl Acad Sci USA* 84:2537–39.

Gintzler, A.R. 1980. Endorphin-mediated increases in pain threshold during pregnancy. *Science* 210:193–95.

Gintzler, A.R., and Bohan, M.C. 1990. Pain thresholds are elevated during pseudopregnancy. *Brain Res* 507:312–16.

Girardot, M.N., and Holloway, F.A. 1984a. Intermittent cold water stress-analgesia in rats: cross-tolerance to morphine. *Pharmacol Biochem Behav* 20:631–33.

———. 1984b. Intermittent cold-water swim analgesia in rats: differential effects of naltrexone. *Physiol Behav* 32:547–55.

Giros, B., Jaber, M., Jones, S., Wightman, R., and Caron, M. 1996. Hyperlocomotion and indifference to cocaine and amphetamine in mice lacking the dopamine transporter. *Nature* 379:606–12.

Gistrak, M.A., Paul, D., Hahn, E.F., and Pasternak, G.W. 1989. Pharmacological actions of a novel mixed opiate agonist-antagonist: naloxone benzoylhydrazone. *J Pharmacol Exp Ther* 251:469–76.

Glass, P., Bloom, M., Kearse, L., Rosow, C., Sebel, P., and Manberg, P. 1997. Bispectral analysis measures sedation and memory effects of propofol, midazolam, isoflurane, and alfentanil in healthy volunteers. *Anesthesiology* 86:836–47.

Glazer, E., and Basbaum, A. 1984. Immunohistochemical localization of leucine-enkephalin in the spinal cord of the cat: enkephalin-containing marginal neurons and pain modulation. *J Comp Neurol* 196:377–89.

Gogas, K.R., and Hough, L.B. 1988. Effects of zolantidine, a brain-penetrating H-2 receptor antagonist, on naloxone-sensitive and naloxone-insensitive forms of footshock-induced analgesia. *Neuropharmacology* 27:357–62.

Gogas, K.R., Hough, L.B., Glick, S.D., and Su, K.L. 1987. Opposing actions of cimetidine on naloxone-sensitive and naloxone-insensitive forms of footshock-induced analgesia. *Brain Res* 370:370–74.

Goldberg, I.E., Rossi, G.C., Letchworth, S.R., Mathis, J.P., Ryan-Moro, J., Leventhal, L., Su, W., Emmel, D., Bolan, E.A., and Pasternak, G.W. 1998. Pharmacological characterization of endomorphin-1 and endomorphin-2 in mouse brain. *J Pharmacol Exp Ther* 286:1007–13.

Goldstein, A., Fischli, W., Lowney, L.I., Hunkapiller, M., and Hood, L. 1981. Porcine pituitary dynorphin: complete amino acid sequence of the biologically active heptadecapeptide. *Proc Natl Acad Sci USA* 74:7219–23.

Goodman, R.R., and Pasternak, G.W. 1985. Visualization of μ-1 opiate receptors in rat brain using a computerized autoradiographic subtraction technique. *Proc Natl Acad Sci USA* 82:6667–71.

Goodman, R.R., Snyder, S.H., Kuhar, M.J., and Young, W.S. 1980. Differentiation of δ and μ opiate receptor localizations by light microscopic autoradiography. *Proc Natl Acad Sci USA* 77:6239–43.

Gordon, W.C., Scobie, S.R., and Frankl, S.E. 1978. Age-related differences in electric shock detection and escape thresholds in Sprague-Dawley albino rats. *Exp Aging Res* 4:23–35.

Goubillon, M.-L., Delaleu, B., Tillet, Y., Caraty, A., and Herbison, A. 1999. Localization of estrogen-receptive neurons projecting to the GnRH neuron-containing rostral preoptic area of the ewe. *Reprod Neuroendocrinol* 70:228–36.

Grattan, D.R., and Selmanoff, M. 1997. Sex differences in the activity of γ-aminobutyric acidergic neurons in the hypothalamus. *Brain Res* 775:244–49.

Grau, J.W., Hyson, R.L., Maier, S.F., Madden, J., and Barchas, J.D. 1981. Long-term stress-induced analgesia and activation of the opiate system. *Science* 213:1409–10.

Greeley, J.D., Le, A.D., Poulos, C.X., and Cappell, H. 1988. Paradoxical analgesia induced by naloxone and naltrexone. *Psychopharmacology* 96:36–39.

Grisel, J.E., Fleshner, M., Watkins, L.R., and Maier, S.F. 1993. Opioid and nonopioid interactions in two forms of stress-induced analgesia. *Pharmacol Biochem Behav* 45:161–72.

Grisel, J.E., Mogil, J.S., Belknap, J.K., and Grandy, D.K. 1996. Orphanin FQ acts as a supraspinal, but not a spinal, anti-opioid peptide. *Neuroreport* 7:2125–29.

Guorraci, F.A., and Kapp, B.S. 1999. An electrophysiological characterization of ventral tegmental area dopaminergic neurons during differential Pavlovian fear conditioning in the awake rabbit. *Behav Brain Res* 99:169–79.

Guimares, A.P.C., and Prado, W.A. 1994. Antinociceptive effects of carbachol microinjected into different portions of the mesencephalic periaqueductal gray matter of the rat. *Brain Res* 647:220–30.

Gysling, K., and Wang, R. 1983. Morphine-induced activation of A10 dopamine neurons in the rat. *Brain Res* 277:119–27.

Hall, J., Duggan, A., Johnson, S., and Morton, C. 1981. Medullary raphe lesions do not reduce descending inhibition of dorsal horn neurones of the cat. *Neurosci Lett* 25:25–29.

Hall, J., Duggan, A., Morton, C., and Johnson, S. 1982. The location of brainstem neurones tonically inhibiting dorsal horn neurones of the cat. *Brain Res* 244:215–22.

Hamberger, B., and Hokfelt, T. 1968. Distribution of noradrenaline nerve terminals in cortical areas of the rat. *Brain Res* 8:125–31.

Hamm, R.J., and Knisely, J.S. 1985. Environmentally-induced analgesia: an age-related decline in an endogenous opioid system. *J Gerontol* 40:268–74.

———. 1986. The analgesia produced by food deprivation in 4-month-old, 14-month-old and 24-month-old rats. *Life Sci* 39:1509–15.

Hamm, R.J., and Lyeth, B.G. 1984. Nociceptive thresholds following food restriction and return to free feeding. *Physiol Behav* 33:499–501.

Hamm, R.J., Knisely, J.S., and Watson, A. 1986. Environmentally-induced analgesia: age-related changes in a hormonally-mediated, nonopioid system. *J Gerontol* 41:336–41.

Hammond, D.L. 1986. Control systems for nociceptive afferent processing. In *Spinal Afferent Processing*, ed. T.L. Yaksh, 363–90. New York: Plenum.

Han, Y., Shaikh, M., and Siegel, A. 1996. Medial amygdaloid suppression of predatory attack behavior in the cat: II. Role of a GABAergic pathway from the medial to the lateral hypothalamus. *Brain Res* 716:72–83.

Han, Y.S., Brutus, M., and Siegel, A. 1996. Medial amygdaloid suppression of predatory attack behavior in the cat: I. Role of a substance P pathway from the medial amygdala to the medial hypothalamus. *Brain Res* 716:59–71.

Harasawa, I., Fields, H., and Meng, I. 2000. Delta opioid receptor mediated actions in the rostral ventromedial medulla on tail flick latency and nociceptive modulatory neurons. *Pain* 85:255–62.

Harlan, R., Shivers, B., Romano, G., Howells, R., and Pfaff, D. 1987. Localization of preproenkephalin mRNA in the rat brain and spinal cord by in situ hybridization. *J Comp Neurol* 258:159–84.

Hasegawa, Y., Kurachi, M., Okuyama, S., Araki, H., and Otomo, S. 1990. 5HT-3 receptor antagonists inhibit the response of K opioid receptors in the morphine-reduced Straub tail. *Eur J Pharmacol* 190:399–401.

Hayes, R.L., Newlon, P.G., Rosecrans, J.A., and Mayer, D.J. 1977. Reduction of stimulation-produced analgesia by lysergic acid diethylamide, a depressor of serotonergic activity. *Brain Res* 122:367–72.

Hayes, R.L., Price, D.D., Bennett, G.J., Wilcox, G.L., and Mayer, D.J. 1978. Differential effects of spinal cord lesions on narcotic and non-narcotic suppression of nociceptive reflexes: further evidence for the physiologic multiplicity of pain modulation. *Brain Res* 155:91–101.

Haywood, S., Simonian, S., van der Beek, E., Bicknell, J., and Herbison, A. 1999. Fluctuating estrogen and progesterone receptor expression in brainstem norepinephrine neurons through the rat estrous cycle. *Endocrinology* 140:3255–63.

Hebb, D.O. 1955. Drives and the C.N.S. *Psychol Rev* 62:243–54.

Heinricher, M., and McGaraughty, S. 1999. Pain-modulating neurons and behavioral state. *Handbook of Behavioral State Control*, 487–503. New York: CRC Press.

Heinricher, M.M., and McGaraughty, S. 1998. Analysis of excitatory amino acid transmission within the rostral ventromedial medulla: implications for circuitry. *Pain* 75:247–55.

Heinricher, M.M., and Roychowdhury, S.M. 1997. Reflex-related activation of putative pain facilitating neurons in rostral ventromedial medulla requires excitatory amino acid transmission. *Neuroscience* 78:1159–65.

Heinricher, M.M., Haws, C.M., and Fields, H.L. 1991. Evidence for GABA-mediated control of putative nociceptive modulating neurons in the rostral ventralmedial medulla: iontophoresis of bicuculline eliminates the off-cell pause. *Somatosens Motor Res* 8:215–25.

Heinricher, M.M., McGaraughty, S., and Farr, D.A. 1999. The role of excitatory amino acid transmission within the rostral ventromedial medulla in the antinociceptive actions of systemically administered morphine. *Pain* 81:57–65.

Heinricher, M.M., Morgan, M.M., and Fields, H.L. 1992. Direct and indirect actions of morphine on medullary neurons that modulate nociception. *Neuroscience* 48:533–43.

Heinricher, M.M., Morgan, M.M., Tortrici, V., and Fields, H.L. 1994. Disinhibition of off-cells and antinociception produced by an opioid action within the rostral ventromedial medulla. *Neuroscience* 63:279–88.

Heisler, L., Chu, H.-M., Brennan, T., Danao, J., Bajwa, P., Parsons, L., and Tecott, L. 1998. Elevated anxiety and antidepressant-like responses in serotonin 5-HT1A receptor mutant mice. *Proc Natl Acad Sci USA* 95:15049–54.

Heller, S.B., Komisaruk, B.R., Gintzler, A.R., and Stracher, A. 1986. Prolongation of vaginal stimulation-produced analgesia by leupeptin, a protease inhibitor. *Ann NY Acad Sci* 467:419–22.

Helmstetter, F.J. 1992. The amygdala is essential for the expression of conditional hypoalgesia. *Behav Neurosci* 106:518–28.

———. 1993. Stress-induced hypoalgesia and defensive freezing are attenuated by application of diazepam to the amygdala. *Pharmacol Biochem Behav* 44: 433–38.

Helmstetter, F.J., and Bellgowan, P.S. 1993. Lesions of the amygdala block conditional hypoalgesia on the tail-flick test. *Brain Res* 612:253–57.

Helmstetter, F.J., Bellgowan, P.S., and Tershner, S.A. 1993. Inhibition of the tail-flick reflex following microinjection of morphine into the amygdala. *Neuroreport* 4:471–74.

Helmstetter, F.J., Bellgowan, P.S.F., and Poore, L.H. 1995. Microinfusions of μ, but not δ or κ opioid agonists into the basolateral amygdala results in inhibition of the tail-flick reflex in pentobarbital-anesthetized rats. *J Pharmacol Exp Ther* 275:381–88.

Helmstetter, F.J., Tershner, S.A., Poore, L.H., and Bellgowan, P.S.F. 1998. Antinociception following opioid stimulation of the basolateral amygdala is expressed through the periaqueductal gray and rostral ventromedial medulla. *Brain Res* 779:104–8.

Henderson, G., and McKnight, A.T. 1997. The orphan opioid receptor and its endogenous ligand—nociceptin/orphanin FQ. *Trends Pharmacol Sci* 18:293–300.

Henderson, L., Krey, K., and Bandler, R. 1998. The ventrolateral periaqueductal gray projects to caudal brainstem depressor regions: a functional-anatomical and physiological study. *Neuroscience* 82:201–21.

Hentall, I., Barbaro, N., and Fields, H. 1991. Spatial and temporal variation of microstimulation thresholds for inhibiting the tail-flick reflex from the rat's rostral medial medulla. *Brain Res* 548:156–62.

Herbison, A. 1997. Noradrenergic regulation of cyclic GnRH secretion. *Rev Reprod* 2:1–6.

Heritage, A., Stumpf, W., Sar, M., and Grant, L. 1980. Brainstem catecholamine neurons are target sites for sex steroid hormones. *Science* 207:1377–79.

Herman, B.H., and Goldstein, A. 1985. Antinociception and paralysis induced by intrathecal dynorphin. *J Pharmacol Exp Ther* 232:27–32.

Hess, G.D., Joseph, J.A., and Roth, G.S. 1981. Effect of age on sensitivity to pain and brain opiate receptors. *Neurobiol Aging* 2:49–55.

Heyman, J.S., Vaught, J.L., Raffa, R.B., and Porreca, F. 1988. Can supraspinal δ opioid receptors mediate antinociception? *Trends Pharmacol Sci* 9:134–38.

Hill, R.G., and Ayliffe, S.J. 1981. The antinociceptive effect of vaginocervical stimulation in the rat is reduced by naloxone. *Pharmacol Biochem Behav* 14:631–32.

Hinde, R. 1960. Energy models of motivation. *Symp Soc Exp Biol* 14:199–213.

———. 1970. *Animal behaviour.* New York: McGraw Hill.

Hobson, J., and Brazier, M., eds. 1980. *The Reticular Formation Revisited: Specifying Function for a Non-Specific System.* New York: Raven.

Hoffman, G.E., Dohanics, J., Watson, R.E., Jr., and Wiegand, S.J. 1996. The hypothalamic ventromedial nucleus sends a met-enkephalin projection to the preoptic area's periventricular zone in the female rat. *Brain Res Mol Brain Res* 36(2):201–10.

Hogervorst, E., Boshuisen, M., Riedel, W., Willeken, C., and Jolles, J. 1998. The effect of hormone replacement therapy on cognitive function in elderly women. *Psychoneuroendocrinology* 24:43–68.

Hokfelt, T. 1967. On the ultrastructural localization of noradrenaline in the central nervous system of the rat. *Zeitschrift fur Zellforschung* 79:110–17.

———. 1968. In vitro studies on central and peripheral monoamine neurons at the ultrastructural level. *Zeitschrift fur Zellforschung* 91:1–74.

———. 1969. Distribution of noradrenaline storing particles in peripheral adrenergic neurons as revealed by electron microscopy. *Acta Physiol Scand* 76:427–40.

Hokfelt, T., Elde, R., Johansson, O., Terenius, L., and Stein, L. 1977. The distribution of enkephalin-immunoreactive cell bodies in the rat central nervous system. *Neurosci Lett* 5:25–31.

Hokfelt, T., Ljungdahl, A., Fuxe, K., and Johansson, O. 1974. Dopamine nerve termianls in the rat limbic cortex: aspects of the dopamine hypothesis of schizophrenia. *Science* 184:177–179.

Holaday, J.H., and Faden, A.I. 1983. Thyrotropin releasing hormone: autonomic effects upon cardiorespiratory function in endotoxic shock. *Regul Pept* 7:111–25.

Holden, J.E., and Proudfit, H.K. 1998. Enkephalin neurons that project to the

A7 catecholamine cell group are located in nuclei that modulate nociception: ventromedial medulla. *Neuroscience* 83:929–47.

Holden, J., Schwartz, E., and Proudfit, H. 1999. Microinjection of morphine in the A7 catecholamine cell group produces opposing effects on nociception that are mediated by α-1-and α-2-adrenoceptors. *Neuroscience* 91:979–90.

Hopkins, E., Spinella, M., Pavlovic, Z.W., and Bodnar, R.J. 1998. Alterations in swim stress-induced analgesia and hypothermia following serotonergic or NMDA antagonists in the rostral ventromedial medulla of rats. *Physiol Behav* 64:219–25.

Horvitz, J. 1997. Burst activity of ventral tegmental dopamine neurons is elicited by sensory stimuli in the awake cat. *Brain Res* 759:251.

Horvitz, J.C. 2000. Mesolimbocortical and nigrostriatal dopamine responses to salient non-reward events. *Neuroscience* 96:651–56.

Hosobuchi, Y., Rossier, J., Bloom, F.E., and Guilleman, R. 1979. Stimulation of human periaqueductal gray for pain relief increases immunoreactive β-endorphin in ventricular fluid. *Science* 203:279–81.

Hough, L.B., Glick, S.D., and Su, K.L. 1985. A role for histamine and histamine H2 receptors in non-opiate footshock-induced analgesia. *Life Sci* 36:859–66.

Hruska, R. 1986. Elevation of striatal dopamine receptors by estrogen: dose and time studies. *J Neurochem* 47:1908–15.

Hruska, R., and Nowak, M. 1988. Estrogen treatment increases the density of $D_1$ dopamine receptors in the rat striatum. *Brain Res* 442:349–50.

Hughes, J., Smith, T., Kosterlitz, H.W., Fothergill, L.A., Morgan, B.A., and Morris, H.R. 1975. Identification of two related penta-peptides from the brain with potent opiate agonist activity. *Nature* 258:577–79.

Hull, C.L. 1943. *Principles of Behavior.* New York: Appleton-Century.

———. 1952. *A Behavior System.* New Haven: Yale University Press.

Hull, E., Du, J., Lorrain, D., and Matuszewich, L. 1997. Testosterone, preoptic dopamine, and copulation in male rats. *Brain Res Bull* 44:327–33.

Hylden, J., Hayashi, H., and Bennett, G. 1986. Lamina I spinomesencephalic neurons in the cat ascend via the dorsolateral funiculi. *Somatosens Res* 4:31–41.

Hyson, R.L., Ashcraft, L.J., Drugan, R.C., Grau, J.W., and Maier, S.F. 1982. Extent and control of shock affects naltrexone sensitivity of stress-induced analgesia and reactivity to morphine. *Pharmacol Biochem Behav* 17:1019–25.

Ingvar, D.H., and Sourander, P. 1970. Destruction of the reticular core of the brain stem: a pathoanatomic follow-up of a case of coma of three years' duration. *Arch Neurol* 23:1–17.

Islam, A.K., Cooper, M.L., and Bodnar, R.J. 1993. Interactions among aging, gender and gonadectomy effects upon morphine antinociception in rats. *Physiol Behav* 54:45–53.

Itzhak, Y., and Pasternak, G.W. 1987. Interaction of [D-Ser2, Leu5]-enkephalin-

Thr6 (DSLET), a relatively selective delta ligand, with mu1 opioid binding sites. *Life Sci* 40:307–11.

Iversen, S. 1998. Behavioral topography in the striatum: differential effects of quinpirole and D-amphetamine microinjections. *Eur J Pharmacol* 362:111–19.

Iwakiri, H., Oka, T., Takakusaki, K., and Mori, S. 1995. Stimulus effects of the medial pontine reticular formation and the mesencephalic locomotor region upon medullary reticulospinal neurons in acute decerebrate cats. *Neurosci Res* 23:47–53.

Iwamoto, E.T. 1989. Antinociception after nicotine administration into the mesopontine tegmentum of rats: evidence for muscarinic actions. *J Pharmacol Exp Ther* 251:412–21.

———. 1991. Characterization of the antinociception induced by nicotine in the pedunculopontine tegmental nucleus and the nucleus raphe magnus. *J Pharmacol Exp Ther* 257:120–33.

Iwamoto, E.T., and Marion, L. 1993a. Adrenergic, serotonergic and cholinergic components of nicotinic antinociception in rats. *J Pharmacol Exp Ther* 265:777–89.

———. 1993b. Characterization of the antinociception produced by intrathecally administered muscarinic agonists in rats. *J Pharmacol Exp Ther* 266:329–38.

Jackson, R.L., Coon, D.J., and Maier, S.F. 1979. Long term analgesia effects of inescapable shock and learned helplessness. *Science* 206:91–94.

Jacobs, B., Abercrombie, E., Fornal, C., Levine, E., Morilak, D., and Stafford, I. 1991. Single-unit and physiological analyses of brain norepinephrine function in behaving animals. *Prog Brain Res* 88:159–65.

Jacquet, Y. 1988. The NMDA receptor: central role in pain inhibition in the periaqueductal gray. *Eur J Pharm* 154:271–76.

Jacquet, Y., and Lajtha, A. 1973. Morphine action at central nervous system sites in rat: analgesia or hyperalgesia depending on site and dose. *Science* 182:490–91.

Jacquet, Y.F., and Lajtha, A. 1974. Paradoxical effects after microinjection of morphine in the periaqueductal gray matter in the rat. *Science* 185:1055–57.

Jacquet, Y.F., Carol, M., and Russell, I.S. 1976. Morphine-induced rotation in naive non-lesioned rats. *Science* 192:261–63.

James, M., MacKenzie, F., Tuohy-Jones, P., and Wilson, C. 1987. Dopaminergic neurons in the zona incerta exert a stimulatory control of gonadotrophin release via $D_1$-dopamine receptors. *Neuroendocrinology* 45:348–55.

Jansen, A., Farkas, E., Sams, J., and Loewy, A. 1998. Local connections between the columns of the periaqueductal gray matter: a case for intrinsic neuromodulation. *Brain Res* 784:329–36.

Jenkins, M. 1928. The effect of segregation on the sex behavior of the white rat as measured by the obstruction method. *Genet Psychol Monogr* 3:455–568.

Jennett, B., and Plum, F. 1973. Persistent vegetative state after brain damage: a syndrome in search of a name. *Lancet* 1:734–37.

Jensen, T.S., and Yaksh, T.L. 1984a. Glutamate-induced analgesia: effects of spinal serotonin, norepinephrine and opioid antagonists. *Adv Pain Res Ther* 9:513–18.

―――. 1984b. Spinal monoamine and opiate systems partly mediate the antinociceptive effects produced by glutamate at brainstem sites. *Brain Res* 321:287–89.

―――. 1986a. I. Comparison of antinociceptive action of morphine in the periaqueductal gray, medial and paramedial medulla in rat. *Brain Res* 363:99–113.

―――. 1986b. II. Examination of spinal monoamine receptors through which brainstem opiate-sensitive systems act in the rat. *Brain Res* 363:114–27.

―――. 1986c. III. Comparison of antinociceptive action of mu and delta opioid receptor ligands in the periaqueductal gray matter, medial and paramedial ventral medulla in the rat as studied by the microinjection technique. *Brain Res* 372:301–12.

Jiang, Q., Takemori, A.E., Sultana, M., Portoghese, P.S., Bowen, W.D., Mosberg, H.I., and Porreca, F. 1991. Differential antagonism of opioid delta antinociception by [D-Ala2, Leu5, Cys6]-enkephalin (DALCE) and naltrindole 5′-isothiocyanate (5′-NTII): evidence for delta receptor subtypes. *J Pharmacol Exp Ther* 257:1069–75.

Jones, B. 1991a. The role of noradrenergic locus coeruleus neurons and neighboring cholinergic neurons of the pontomesencephalic tegmentum in sleep-wake states. *Prog Brain Res* 88:533–43.

―――. 1991b. Noradrenergic locus coeruleus neurons: their distant connections and their relationship to neighboring (including cholinergic and GABAergic) neurons of the central gray and reticular formation. *Prog Brain Res* 88:15–30.

―――. 1993. The organization of central cholinergic systems and their functional importance in sleep-waking states. *Prog Brain Res* 98:61–71.

Jones, B., and Cuello, A. 1989. Afferents to the basal forebrain cholinergic cell area from pontomesencephalic—catecholamine, serotonin and acetylcholine—neurons. *Neuroscience* 31:37–61.

Jones, B., and Yang, T. 1985. The efferent projections from the reticular formation and the locus coeruleus studied by anterograde and retrograde axonal transport in the rat. *J Comp Neurol* 242:56–92.

Jones, B., Harper, S., and Halaris, A. 1977. Effects of locus coeruleus lesions upon cerebral monoamine content, sleep-wakefulness states and the response to amphetamine in the cat. *Brain Res* 124:473–96.

Jones, S., Gainetdinov, R., Jaber, M., Giros, B., Wightman, R., and Caron, M. 1998. Profound neuronal plasticity in response to inactivation of the dopamine transporter. *Proc Natl Acad Sci USA* 95:4029–34.

Jones, S.L., and Blair, R.W. 1995. Noxious heat-evoked Fos-like immunoreactivity in the rat medulla, with emphasis on the catecholamine cell groups. *J Comp Neurol* 354:410–22.

Jorgenson, K., Kow, L., and Pfaff, D. 1989. Histamine excites arcuate neurons in vitro through H1 receptors. *Brain Res* 502:171–79.

Julius, D. 1998. Serotonin receptor knockouts: a moody subject. *Proc Natl Acad Sci USA* 95:15153–54.

Justice, A., and de Wit, H. 2000. Acute effects of estradiol pretreatment on the response to d-amphetamine in women. *Neuroendocrinology* 71:51–59.

Kalivas, P.W., Gau, B.A., Nemeroff, C.B., and Prange, A.J. 1982. Antinociception after microinjection of neurotensin into the central amygdaloid nucleus of the rat. *Brain Res* 243:279–86.

Kangawa, K., Minamino, N., Chino, N., Sakakibara, S., and Matsuo, H. 1981. The complete amino acid sequence of alpha-neo-endorphin. *Biochem Biophys Res Commun* 99:871–78.

Kaplan, H., and Fields, H. 1991. Hyperalgesia during acute opioid abstinence: evidence for a nociceptive facilitating function of the rostral ventromedial medulla. *J Neurosci* 11:1433–39.

Kator, A., and Sakuma, Y. 2000. Neuronal activity in female rat preoptic area associated with sexually motivated behavior. *Brain Res* 862:90–102.

Kauser, K., and Rubanyi, G. 1998. Estrogen and nitric oxide in vasculature. *Curr Opin Endocrinol Diabetes* 6:230–37.

Kavaliers, M. 1988. Brief exposure to a natural predator, the short-tailed weasel, induces benzodiazepine-sensitive analgesia in white-footed mice. *Physiol Behav* 43:187–93.

Kavaliers, M., and Choleris, E. 1997. Sex differences in N-methyl-D-aspartate involvement in κ opioid and non-opioid predator-induced analgesia in mice. *Brain Res* 768:30–36.

Kavaliers, M., and Colwell, D.D. 1991. Sex differences in opioid and non-opioid mediated predator-induced analgesia in mice. *Brain Res* 568:173–77.

Kavaliers, M., and Galea, L.A.M. 1995. Sex differences in the expression and antagonism of swim stress-induced analgesia in deer mice vary with the breeding season. *Pain* 63:327–34.

Kavaliers, M., and Innes, D.G.L. 1987a. Stress-induced analgesia and activity in deer mice: sex and population differences. *Brain Res* 425:49–56.

Kavaliers, M., and Innes, D. 1987b. Sex differences in magnetic field inhibition of morphine-induced responses of wild deer mice, *Peromyscus maniculatus triangularis. Physiol Behav* 40:559–62.

Kavaliers, M., and Innis, D.G.L. 1987c. Sex and day/night differences in opiate-induced responses of insular wild deer mice, *Peromyscus maniculatis triangularis. Pharmacol Biochem Behav* 27:477–82.

Kavaliers, M., and Innes, D.G.L. 1988. Novelty-induced opioid analgesia in deer mice (*Peromyscus maniculatus*): sex and population differences. *Behav Neural Biol* 49:54–60.

————. 1990. Developmental changes in opiate-induced analgesia in deer mice: sex and population differences. *Brain Res* 16:326–31.

Kavaliers, M., and Innes, D.G.L. 1992a. Sex differences in the effects of Tyr-MIF-1 on morphine and stress-induced analgesia. *Peptides* 13:1295–97.

Kavaliers, M., and Innes, D. 1992b. Sex differences in the effects of neuropeptide FF and IgG from neuropeptide FF on morphine-and stress-induced analgesia. *Peptides* 13:603–7.

Kavaliers, M., and Innes, D. 1993a. Sex differences in naloxone-and Tyr-MIF-1-induced hypoalgesia. *Peptides* 14:1001–4.

Kavaliers, M., and Innes, D.G. 1993b. Sex differences in the antinociceptive effects of the enkephalinase inhibitor, SCH 34826. *Pharmacol Biochem Behav* 46:777–80.

Kavaliers, M., and Wiebe, J.P. 1987. Analgesic effects of the progesterone metabolite, 3α-hydroxy-5α-pregnan-20-one and possible modes of action. *Brain Res* 415:393–98.

Kavaliers, M., Colwell, D., and Choleris, E. 1998. Sex differences in opioid and N-methyl-D-aspartate mediated non-opioid biting fly exposure induced analgesia in deer mice. *Pain* 77:163–71.

Kavaliers, M., Hirst, M., and Teskey, G.C. 1983. Aging, opioid analgesia and the pineal gland. *Life Sci* 32:2279–87.

Keay, K.A., Crowfoot, L.J., Floyd, N.S., Henderson, L.A., Christie, M.J. and Bandler, R. 1997. Cardiovascular effects of microinjections of opioid agonists into the 'Depressor Region' of the ventrolateral periaqueductal gray region. *Brain Res* 762:61–71.

Kelly, D.D., Silverman, A.J., Glusman, M., and Bodnar, R.J. 1993. Characterization of pituitary mediation of stress-induced antinociception in rats. *Physiol Behav* 53:769–75.

Kepler, K.L., and Bodnar, R.J. 1988. Yohimbine potentiates cold-water swim analgesia: re-evaluation of a noradrenergic role. *Pharmacol Biochem Behav* 29:83–88.

Kepler, K.L., Kest, B., Kiefel, J.M., Cooper, M.L., and Bodnar, R.J. 1989. Roles of gender, gonadectomy and estrous phase in the analgesic effects of intracerebroventricular morphine in rats. *Pharmacol Biochem Behav* 34:119–27.

Kepler, K.L., Standifer, K.M., Paul, D., Pasternak, G.W., Kest, B., and Bodnar, R.J. 1991. Differential gender effects upon central opioid analgesia. *Pain* 45:87–95.

Kessler, R., McGonagle, K., Zhao, S., Nelson, C., Hughes, M., Eshleman, S., Wittchen, H., and Kindler, K. 1994. Lifetime and 12-month prevalence of DSM-III-R psychiatric disorders in the United States. Results from the National Comorbidity Survey. *Arch Gen Psychiatry* 51:8–19.

Kest, B., Beczkowska, I., Franklin, S.O., Lee, C.E., Mogil, J.S., and Inturrisi, C.E. 1998a. Differences between delta opioid receptor antinociception, binding and mRNA levels between BALB/c and CXBK mice. *Brain Res* 805:131–37.

Kest, B., Jenab, S., Brodsky, M., Sadowsky, B., Belknap, J.K., Mogil, J.S., and Inturrisi, C.E. 1999. µ and δ opioid receptor analgesia, binding density, and mRNA levels in mice selectively bred for high and low analgesia. *Brain Res* 816:381–89.

Kest, B., Lee, C.E., Mogil, J.S., and Inturrisi, C.E. 1997. Blockade of morphine supersensitivity by an antisense oligodeoxynucleotide targeting the delta opioid receptor (DOR-1). *Life Sci* 60:PL155–59.

Kest, B., McLemore, G.L., Sadowski, B., Mogil, J.S., Belknap, J.K., and Inturrisi, C.E. 1998b. Acute morphine dependence in mice selectively bred for high and low analgesia. *Neurosci Lett* 256:120–22.

Kest, B., Orlowski, M., and Bodnar, R.J. 1991. Increases in opioid-mediated swim antinociception following endopeptidase 24.15 inhibition. *Physiol Behav* 50:843–45.

———. 1992. Endopeptidase 24.15 inhibition and opioid antinociception. *Psychopharmacology* 106:408–16.

Kest, B., Wilson, S.G., and Mogil, J.S. 1999. Sex differences in supraspinal morphine analgesia are dependent on genotype. *J Pharmacol Exp Ther* 289:1370–75.

Khachaturian, H., Lewis, M.E., Schaffer, K.H.M., and Watson, S. 1985. Anatomy of the CNS opioid systems. *Trends Neurosci* 1:10–19.

Khateb, A., Fort, P., Pegna, A., Jones, B., and Muhlethaler, M. 1995. Cholinergic nucleus basalis neurons are excited by histamine in vitro. *Neuroscience* 69:495–506.

Kiefel, J.M., and Bodnar, R.J. 1992. Roles of gender and gonadectomy in pilocarpine and clonidine analgesia in rats. *Pharmacol Biochem Behav* 41:153–58.

Kiefel, J.M., Cooper, M.L., and Bodnar, R.J. 1992. Serotonin receptor subtype antagonists in the medial ventral medulla inhibit mesencephalic opiate analgesia. *Brain Res* 597:331–38.

Kiefel, J.M., Paul, D., and Bodnar, R.J. 1989. Reduction in opioid and nonopioid forms of swim analgesia by 5HT-2 antagonists. *Brain Res* 500:231–40.

Kiefel, J.M., Rossi, G.C., and Bodnar, R.J. 1993. Medullary µ and δ opioid receptors modulate mesencephalic morphine analgesia in rats. *Brain Res* 624:151–60.

Kieffer, B., Befort, K., Gareiaux-Ruff, C., and Hirth, C. 1992. The δ-opioid receptor: isolation of cDNA by expression cloning and pharmacological characterization. *Proc Natl Acad Sci USA* 89:12048–52.

Kiehn, O., Harris-Warrick, R., Jordan, L., Hultborn, H., and Kudo, N., eds. 1998. *Neuronal Mechanisms for Generating Locomotor Activity*. Annals of the New York Academy of Sciences. New York.

Kimura, S., Lewis, R.V., Stern, A.S., Rossier, J., Stein, S., and Undenfriend, S. 1980. Probable precursors of (leu) and (met)-enkephalin in adrenal medulla: peptides of 3–5 kilodaltons. *Proc Natl Acad Sci USA* 77:1681–85.

King, M.A., Rossi, G.C., Chang, A.H., Williams, L., and Pasternak, G.W. 1997.

Spinal analgesic activity of orphanin FQ/nociceptin and its fragments. *Neurosci Lett* 223:113–16.

Kirchgessner, A.L., Bodnar, R.J., and Pasternak, G.W. 1982. Naloxazone and pain-inhibitory systems: evidence for a collateral inhibition model. *Pharmacol Biochem Behav* 17:1175–79.

Kitchen, I., Slowe, S.J., Matthes, H.W., and Kieffer, B. 1997. Quantitative autoradiographic mapping of $\mu$-, $\delta$- and $\kappa$-opioid receptors in knockout mice lacking the $\mu$-opioid receptor gene. *Brain Res* 778:73–88.

Klamt, J.G., and Prado, W.A. 1991. Antinociception and behavioral changes induced by carbachol microinjected into identified sites of the rat brain. *Brain Res* 549:9–18.

Kolesnikov, Y.A., Pick, C.G., and Pasternak, G.W. 1992. NG-Nitro-L-Arginine prevents morphine tolerance. *Eur J Pharmacol* 221:399–400.

Kolesnikov, Y.A., Pick, C.G., Ciszewska, G., and Pasternak, G.W. 1993. Blockade of tolerance to morphine but not kappa opioids by a nitric oxide synthase inhibitor. *Proc Natl Acad Sci USA* 90:5162–66.

Kolesnikov, Y., Jain, S., Wilson, R., and Pasternak, G. 1998. Lack of morphine and enkephalin tolerance in 129/SvEv mice: evidence for a NMDA receptor defect. *J Pharmacol Exp Ther* 284:455–59.

Kolmodin, G., and Skogland, C. 1960. Analysis of spinal interneurons activated by tactile and nociceptive stimulation. *Acta Physiol Scand* 50:337–55.

Komisaruk, B. 1974. Neural and hormonal interactions in the reproductive behavior of female rats. *Adv Behav Biol* 11:97–129.

Komisaruk, B.R., and Whipple, B. 1986. Vaginal stimulation-produced analgesia in rats and women. *Ann NY Acad Sci* 467:30–39.

Kordon, C., Drouva, S., Martinez de la Escalera, G., and Weiner, R. 1994. Role of classic and peptide neuromediators in the neuroendocrine regulation of luteinizing hormone and prolactin. In *The Physiology of Reproduction,* vol. 1, ed. E. Knobil and J. Neill, 1621–82. New York: Raven.

Kordower, J.H., and Bodnar, R.J. 1984. Vasopressin analgesia: specificity of action and non-opioid effects. *Peptides* 5:747–56.

Kordower, J.H., Sikorszky, V., and Bodnar, R.J. 1982. Central antinociceptive effects of lysine vasopressin and an analogue. *Peptides* 3:613–17.

Kow, L., Weesner, G., and Pfaff, D. 1992. $\alpha$1-adrenergic agonists act on the ventromedial hypothalamus to cause neuronal excitation and lordosis facilitation: electrophysiological and behavioral evidence. *Brain Res* 588:237–45.

Kow, L.-M., and Pfaff, D. 1973. Effects of estrogen treatment on the size of receptive field and response threshold of pudendal nerve in the female rat. *Neuroendocrinology* 13:299–313.

———. 1988. Transmitter and peptide actions on hypothalamic neurons in vitro: implications for lordosis. *Brain Res Bull* 20:857–61.

Kow, L.M., and Pfaff, D.W. 1982. Responses of medullary reticulospinal and other reticular neurons to somatosensory and brainstem stimulation in anesthetized or freely moving rats with or without estrogen treatment. *Exp Brain Res* 47:191–202.

Kow, L.-M., Commons, K., Ogawa, S., and Pfaff, D. 1998. Potentiation of the excitatory action of NMDA in ventrolateral periaqueductal gray (vlPAG) by the m-opioid receptor (MOR) agonist, DAMGO. *Soc Neurosci Abstr* 28:1133.

Kow, L.-M., Johnson, A., Ogawa, S., and Pfaff, D. 1991. Electrophysiological actions of oxytocin on hypothalamic neurons, in vitro: neuropharmacological characterization and effects of ovarian steroids. *Neuroendocrinology* 54:526–35.

Kramer, E., and Bodnar, R.J. 1986a. Age-related decrements in morphine analgesia: a parametric analysis. *Neurobiol Aging* 7:185–91.

———. 1986b. Age-related decrements in the analgesic response to cold-water swims. *Physiol Behav* 36:875–80.

Kramer, E., Sperber, E.S., and Bodnar, R.J. 1985. Age-related decrements in the analgesic and hyperphagic responses to 2-deoxy-D-glucose. *Physiol Behav* 35:929–34.

Krieger, M.S., Conrad, L.C.A., and Pfaff, D.W. 1979. An autoradiographic study of the efferent connections of the ventromedial nucleus of the hypothalamus. *J Comp Neurol* 183:785–816.

Kristal, M.B., Thompson, A.C., Abbott, P., DiPirro, J.M., Ferguson, E.J., and Doerr, J.C. 1990. Amniotic-fluid ingestion by parturient rats enhances pregnancy-mediated analgesia. *Life Sci* 46:693–98.

Krzanowska, E., and Bodnar, R. 2000a. Analysis of sex and gonadectomy differences in β-endorphin antinociception elicited from the ventrolateral periaqueductal gray in rats. *Eur J Pharmacol* 392:157–61.

———. 2000b. Sex differences in locomotor activity following β-endorphin in the ventrolateral periaqueductal gray. *Physiol Behav* 68:595–98.

Krzanowska, E., and Bodnar, R.J. 1999. Morphine antinociception elicited from the ventrolateral periaqueductal gray is sensitive to sex and gonadectomy differences in rats. *Brain Res* 821:224–30.

Krzanowska, E., Znamensky, V., Ragnauth, A., Ogawa, S., Pfaff, D., and Bodnar, R. 2000a. Neonatal gonadectomy alters patterns of sex differences in morphine analgesia elicited from the periaqueductal gray in rats. *Soc Neurosci Abstr* 26, in press.

Krzanowska, E., Znamensky, V., Wilk, S., and Bodnar, R. 2000b. Antinociceptive and behavioral activation responses elicited by D-Pro2-endomorphin-2 in the ventrolateral periaqueductal gray are sensitive to sex differences. *Peptides* 21:705–15.

Kulling, P., Frischnecht, H.R., Pasi, A., Waser, P.G., and Siegfried, B. 1987. Effects of repeated as compared to single aggressive confrontation on nociception and defense behavior in C57BL/6 and DBA/2 mice. *Physiol Behav* 39:599–605.

———. 1988. Social conflict-induced changes in nociception and β-endorphin-like immunoreactivity in pituitary and discrete brain areas of C57BL/6 and DBA/2 mice. *Brain Res* 450:237.

Lai, J., Bilsky, E.J., Bernstein, R.N., Rothman, R.B., Pasternak, G.W., and Porreca, F. 1994a. Antisense oligodeoxynucleotide to the cloned delta opioid receptor selectively inhibits supraspinal, but not spinal antinociceptive effects of [D-Ala2,Glu4]deltorphin. *Regul Pept* 54:159–60.

Lai, J., Bilsky, E.J., Rothman, R.B., and Porreca, F. 1994b. Treatment with antisense oligodeoxynucleotide to the opioid delta receptor selectively inhibits delta2-agonist antinociception. *Neuroreport* 5:1049–52.

Lai, J., Riedl, M., Stone, L.S., Arvidsson, U., Bilsky, E.J., Wilcox, G.L., Elde, R., and Porreca, F. 1996. Immunofluorescence analysis of antisense oligodeoxynucleotide-mediated knock-down of the mouse δ opioid receptor in vitro and in vivo. *Neurosci Lett* 213:205–8.

Langemark, M., Bach, F., Jensen, T., and Olesen, J. 1993. Decreased nociceptive flexion reflex threshold in chronic tension-type headache. *Arch Neurol* 50:1061–64.

Langemark, M., Jensen, K., Jensen, T., and Olesen, J. 1989. Pressure pain thresholds and thermal nociceptive thresholds in chronic tension-type headache. *Pain* 38:203–10.

Langub, M.C., and Watson, R.E.J. 1992. Estrogen receptive neurons in the preoptic area of the rat are postsynaptic targets of a sexually dimorphic enkephalinergic fiber plexus. *Brain Res* 573:61–69.

Lauber, A.H., and Pfaff, D.W. 1990. Estrogen regulation of mRNAs in the brain and relationship to lordosis behavior. *Curr Top Neuroendocrinol* 10:115–47.

Lauber, A.H., Romano, G.J., Mobbs, C.V., Howells, R.D., and Pfaff, D.W. 1990. Estradiol induction of proenkephalin messenger RNA in hypothalamus: dose-response and relation to reproductive behavior in the female rat. *Mol Brain Res* 8:47–54.

Lautenbacher, S., and Rollman, G. 1993. Sex differences in responsiveness to painful and non-painful stimuli are dependent upon the stimulation method. *Pain* 53:255–64.

Leander, J.D., Gesellchen, P.D., and Mendelsohn, L.G. 1986. Comparison of two penicillamine-containing enkephalins: μ, not δ activity produces analgesia. *Neuropeptides* 8:119–25.

LeBars, D., Menetrey, D., Conseiller, C., and Besson, J.M. 1975. Depressive effect of morphine upon lamina V cell activities in the dorsal horn of the spinal cat. *Brain Res* 98:261–77.

Lechner, S., Curtis, A., Brons, R., and Valentino, R. 1997. Locus coeruleus activation by colon distention: role of corticotropin-releasing factor and excitatory amino acids. *Brain Res* 756:114–24.

Legan, S., and Callahan, W. 1999. Suppression of tonic luteinizing hormone secretion and norepinephrine release near the GnRH neurons by estradiol in ovariectomized rats. *Reprod Neuroendocrinol* 70:237–45.

Le Moal, M., and Simon, H. 1991. Mesocorticolimbic dopaminergic network: functional and regulatory roles. *Physiol Rev* 71:155–210.

Letchworth, S., Mathis, J., Rossi, G., Bodnar, R., and Pasternak, G. 2000. Autoradiographic localization of 125-I[Tyr14]Orphanin FQ/Nociceptin and 125-I[Tyr10]Orphanin FQ/Nociceptin1-11 binding sites in rat brain. *J Comp Neurol* 423:319–29.

Leung, C., and Mason, P. 1998. Physiological survey of medullary raphe and magnocellular reticular neurons in the anesthetized rat. *J Neurophysiol* 80: 1630–46.

Lewis, J.W., Cannon, J.T., and Liebeskind, J.C. 1980. Opioid and nonopioid mechanisms of stress analgesia. *Science* 208:623–25.

———. 1983. Involvement of central muscarinic cholinergic mechanisms in opioid stress analgesia. *Brain Res* 270:289–93.

Lewis, J.W., Sherman, J.E., and Liebeskind, J.C. 1981. Opioid and nonopioid stress analgesia: assessment of tolerance and cross-tolerance with morphine. *J Neurosci* 1:358–63.

Lewis, J.W., Terman, G.W., Watkins, L.R., Mayer, D.J., and Liebeskind, J.C. 1983. Opioid and nonopioid mechanisms of footshock-induced analgesia: role of the spinal dorsolateral funiculus. *Brain Res* 267:139–44.

Lewis, J.W., Tordoff, M.G., Sherman, J.E., and Liebeskind, J.C. 1982. Adrenal medullary enkephalin-like peptides may mediate opioid stress analgesia. *Science* 217:557–59.

Lewis, M.E., Khachaturian, H., and Watson, S.J. 1985. Combined autoradiographic-immunocytochemical analysis of opioid receptors and opioid peptide neuronal systems in brain. *Peptides* 6:37–47.

Lewis, V.A., and Gebhart, G.F. 1977. Evaluation of the periaqueductal central gray as a morphine-specific locus of action and examination of morphine-induced and stimulation-produced analgesia at coincident PAG loci. *Brain Res* 124:283–303.

Liebeskind, J.C., Guilbaud, G., Besson, J.M., and Oliveras, J.L. 1973. Analgesia from electrical stimulation of the periaqueductal gray matter in the cat. *Brain Res* 50:441–46.

Lindsley, D.B. 1951. Emotion. In *Handbook of Experimental Psychology,* ed. S.S. Stevens, 473–516. New York: Wiley.

———. 1960. Attention, consciousness, sleep and wakefulness. In *Handbook of Physiology: Neurophysiology III,* ed. J. Field, 1553–93. Washington, D.C.: American Physiological Society.

Lindsley, D.B., Bowden, J.W., and Magoun, H.W. 1949. Effect upon the EEG of acute injury to the brain stem activating system. *EEG Clin Neurophysiol* 1: 475–86.

Lindsley, D.B., Schreiner, L.H., Knowles, W.B., and Magoun, H.W. 1950. Behavioral and EEG changes following chronic brain stem lesions in the cat. *EEG Clin Neurophysiol* 1:483–98.

Ling, G.S.F., Macleod, J.M., Lee, S., Lockhart, S., and Pasternak, G.W. 1984. Separation of morphine analgesia from physical dependence. *Science* 226:462–64.

Ling, G.S.F., Simantov, R., Clark, J.A., and Pasternak, G.W. 1986. Naloxonazine action in vivo. *Eur J Pharmacol* 129:33–38.

Lipa, S., and Kavaliers, M. 1990. Sex differences in the inhibitory effects of the NMDA antagonist, MK-801, on morphine and stress-induced analgesia. *Brain Res Bull* 24:627–30.

Liu, N., and Gintzler, A. 1999. Gestational and ovarian sex steroid antinociception: relevance of uterine afferent and spinal α2-noradrenergic activity. *Pain* 83:359–68.

———. 2000. Prolonged ovarian sex steroid treatment of male rats produces antinociception: identification of sex-based divergent analgesic mechanisms. *Pain* 85:273–81.

Ljungberg, T., Apicerla, P., and Schultz, W. 1991. Responses of monkey midbrain dopamine neurons during delayed alternation performance. *Brain Res* 567:337–41.

Llewelyn, M.B., Azami, J., and Roberts, M.H.T. 1983. Effects of 5-hydroxytryptamine applied into nucleus raphe magnus on nociceptive thresholds and neuronal firing rate. *Brain Res* 258:59–68.

———. 1984. The effect of modification of 5-hydroxytryptamine function in nucleus raphe magnus on nociceptive threshold. *Brain Res* 306:165–70.

Loh, H.H., Tseng, L.F., Wei, E., and Li, C.H. 1976. Beta-endorphin is a potent analgesic agent. *Proc Natl Acad Sci USA* 73:2895–98.

London, E.D., Walter, S.B., and Wamsley, J.K. 1985. Autoradiographic localization of [³H]nicotine binding sites in the rat brain. *Neurosci Lett* 53:179–84.

Long, J.B., Petras, J.M., Mobley, W.C., and Holaday, J.W. 1988. Neurological dysfunction after intrathecal injection of dynorphin A1–13 in the rat. II Nonopioid mechanisms mediate loss of motor, sensory and autonomic function. *J Pharmacol Exp Ther* 246:1167–74.

Lord, J.A.H., Waterfield, A.A., Hughes, J., and Kosterlitz, H. 1977. Endogenous opioid peptides: multiple agonists and receptors. *Nature* 267:495–99.

Lorenz, K. 1950. The comparative method in studying innate behaviour patterns. *Symp Soc Exp Biol* 4:221–68.

Lovick, T. 1985. Ventrolateral medullary lesions block the antinociceptive and cardiovascular responses elicited by stimulating the dorsal periaqueductal gray matter in rats. *Pain* 21:241–52.

Luine, V.N., Grattan, D.R., and Selmanoff, M. 1997. Gonadal hormones alter hypothalamic GABA and glutamate levels. *Brain Res* 747(1):165–68.

Luine, V.N., Wu, V., Hoffman, C.S., and Renner, K.J. 1999. GABAergic regulation of lordosis: influence of gonadal hormones on turnover of GABA and interaction of GABA with 5-HT. *Neuroendocrinology* 69(6):438–45.

Lundberg, A., Malmgren, K., and Schomburg, E. 1987. Reflex pathways from group II muscle afferents. 3. Secondary spindle afferents and the FRA: a new hypothesis. *Exp Brain Res* 65:294–306.

Lupica, C. 1995. δ and μ Enkephalins inhibit spontaneous GABA-mediated IP-SPs via a cyclic AMP-independent mechanism in the rat hippocampus. *J Neurosci* 15:737–49.

Ma, Q.-P., Shi, Y.-S., and Han, J.-S. 1992. Further studies on interactions between periaqueductal gray, nucleus accumbens and habenula in antinociception. *Brain Res* 583:292–95.

MacLennan, A.J., Drugan, R.C., and Maier, S.F. 1983. Long-term stress-induced analgesia blocked by scopolamine. *Psychopharmacology* 80:267–68.

MacLennan, A.J., Drugan, R.C., Hyson, R.L., Maier, S.F., Madden, J., and Barchas, J.D. 1982. Corticosterone: a critical factor in an opioid form of stress-induced analgesia. *Science* 215:1530–32.

Madden, J., Akil, H., Patrick, R.L., and Barchas, J.D. 1977. Stress-induced parallel changes in central opioid level and pain responsiveness. *Nature* 265:358–60.

Maeda, K., Nagatani, S., Estacio, M., and Tsukamura, H. 1996. Novel estrogen feedback sites associated with stress induced suppression of luteinizing hormone secretion in female rats. *Cell Mol Neurobiol* 16:311–24.

Magoun, H.W. 1958. *The Waking Brain*. Springfield, Ill.: Charles C Thomas.

Maier, S.F., Davies, S., Grau, J.W., Jackson, R.L., Morrison, D., Moye, T., Madden, J., and Barchas, J.D. 1980. Opiate antagonists and the long-term analgesia reaction induced by inescapable shock. *J Comp Physiol Psychol* 94:1172–84.

Maier, S.F., Drugan, R.C., and Grau, J.W. 1982. Controllability, coping behavior and stress-induced analgesia in the rat. *Pain* 12:47–56.

Maier, S.F., Sherman, J.E., Lewis, J.W., Terman, G.W., and Liebeskind, J.C. 1983. The opioid-nonopioid nature of stress-induced analgesia and learned helplessness. *J Exp Psychol: Anim Behav Processes* 9:80–91.

Mains, R.E., Eipper, B.A., and Ling, N. 1977. Common precursor to corticotropins and endorphin. *Proc Natl Acad Sci USA* 1974:3014–18.

Maixner, W., and Humphrey, C. 1993. Gender differences in pain and cardiovascular responses to forearm ischemia. *Clin J Pain* 2:16–25.

Maixner, W., Fillingim, R., Booker, D., and Sigurdsson, A. 1995. Sensitivity of patients with painful temporomandibular disorders to experimentally evoked pain. *Pain* 63:341–51.

Maixner, W., Fillingim, R., Sigurdsson, A., Kincaid, S., and Silva, S. 1998. Sensitivity of patients with temporomandibular disorders to experimentally evoked pain: evidence for altered temporal summation of pain. *Pain* 76:71–81.

Majewska, M., Harrison, N., Schwartz, R., Barker, J., and Paul, S. 1986. Steroid hormone metabolites are barbiturate-like modulators of the GABA receptor. *Science* 232:1004–7.

Maloney, K., Mainville, L., and Jones, B. 1998. Differential c-Fos expression in cholinergic, monoaminergic, and GABA-ergic cell groups of the pontomesencephalic tegmentum after paradoxical sleep deprivation and recovery. *J Neurosci* 19:3057–72.

Malsbury, C., Kow, L.-M., and Pfaff, D. 1977. Effects of medial hypothalamic lesions on the lordosis response and other behaviors in female golden hamsters. *Physiol Behav* 19:223–37.

Manning, B.H., and Mayer, D.J. 1995a. The central nucleus of the amygdala contributes to the production of morphine antinociception in the formalin test. *Pain* 63:141–52.

———. 1995b. The central nucleus of the amygdala contributes to the production of morphine antinociception in the tail-flick test. *J Neurosci* 15:8199–8213.

Mansour, A., Fox, C.A., Akil, H., and Watson, S.J. 1995. Opioid-receptor mRNA expression in the rat CNS: anatomical and functional implications. *Trends Neurosci* 18:22–29.

Mansour, A., Khachaturian, H., Lewis, M.E., Akil, H., and Watson, S.J. 1988. Anatomy of CNS opioid receptors. *Trends Neurosci* 11:308–14.

Mantyh, P.W. 1983a. Connections of midbrain periaqueductal gray in the monkey. I. Ascending efferent projections. *J Neurophysiol* 49(3):567–81.

———. 1983b. Connections of midbrain periaqueductal gray in the monkey. II. Descending efferent projections. *J Neurophysiol* 49(3):582–94.

Marcus, D. 1995. Interrelationships of neurochemicals, estrogen and recurring headache. *Pain* 62:129–41.

Marek, P., Ben-Eliyahu, S., Gold, M.S., and Liebeskind, J.C. 1991a. Excitatory amino acid antagonists (kynurenic acid and MK-801) attenuate the development of morphine tolerance in the rat. *Brain Res* 547:77–81.

Marek, P., Ben-Eliyahu, S., Vaccarino, A.L., and Liebeskind, J.C. 1991b. Delayed application of MK-801 attenuates development of morphine tolerance in rats. *Brain Res* 558:163–65.

Marek, P., Mogil, J.S., Sternberg, W.F., Panocka, I., and Liebeskind, J.C. 1992. N-methyl-D-aspartic (NMDA) receptor antagonist MK-801 blocks non-opioid stress-induced analgesia. II. Comparison across three swim stress paradigms in selectively bred rats. *Brain Res* 578:197–203.

Marek, P., Page, G.G., Ben-Eliyahu, S., and Liebeskind, J.C. 1991c. NMDA receptor antagonist MK-801 blocks non-opioid stress-induced analgesia. I. Comparison of opiate receptor-deficient and opiate receptor-rich strains of mice. *Brain Res* 551:293–96.

Marek, P., Ponocka, I., and Hartmann, G. 1982. Enhancement of stress-induced analgesia in adrenalectomized mice: its reversal by dexamethasone. *Pharmacol Biochem Behav* 16:403–5.

Marek, P., Yirmiya, R., and Liebeskind, J. 1988. Strain differences in the magnitude of swimming-induced analgesia in mice correlate with brain opiate receptor concentration. *Brain Res* 447:188–90.

Marek, P., Yirmiya, R., and Liebeskind, J. C. 1990. Genetic influences on brain stimulation–produced analgesia in mice. II. Correlation with brain opiate receptor concentration. *Brain Res* 507:155–7.

Marek, P., Yirmiya, R., Ponocka, I., and Liebeskind, J. C. 1989. Genetic influences on brain stimulation–produced analgesia in mice. I. Correlation with stress-induced analgesia. *Brain Res* 489(1):182–4.

Marks, H.E., and Hobbs, S.H. 1972. Changes in stimulus reactivity following gonadectomy in male and female rats of different ages. *Physiol Behav* 8:113–19.

Marrocco, R., and Davidson, M. 1996. Neurochemistry of attention.In *Mechanisms of Attention*, ed. R. Parasuraman, 35–50. Cambridge, Mass.: MIT Press.

Martin, W.R., Eades, C.G., Thompson, J.A., Huppler, R.E., and Gilbert, P.E. 1976. The effects of morphine-and nalorphine-like drugs in the nondependent and morphine-dependent chronic spinal dog. *J Pharmacol Exp Ther* 197:517–32.

Martin-Iverson, M., and Iversen, S. 1989. Day and night locomotor activity effects during administration of (+)-amphetamine. *Pharmacol Biochem Behav* 34:465–71.

Martin-Schild, S., Gerall, A.A., Kastin, A.J., and Zadina, J.E. 1999. Differential distribution of endomorphin-1 and endomorphin-2-like immunoreactivities in the CNS of the rodent. *J Comp Neurol* 405:450–71.

Martin-Schild, S., Zadina, J.E., Gerall, A.A., Vigh, S., and Kastin, A.J. 1997. Localization of endomorphin-2-like immunoreactivity in the rat medulla and spinal cord. *Peptides* 18:1641–49.

Mason, P. 1997. Physiological identification of pontomedullary serotonergic neurons in the rat. *J Neurophysiol* 77:1087–98.

Mason, P., Back, S., and Fields, H. 1992. A confocal laser microscopic study of enkephalin-immunoreactive appositions onto physiologically identified neurons in the rostral ventromedial medulla. *J Neurosci* 12:4023–36.

Maswood, N., Caldarola-Pastuszka, M., and Uphouse, L. 1997. 5-HT3 receptors in the ventromedial hypothalamus and female sexual behavior. *Brain Res* 769:13–20.

Mathews, D., and Edwards, D. 1977. Involvement of the ventromedial and anterior hypothalamic nuclei in the hormonal introduction of receptivity in the female rat. *Physiol Behav* 19:319–26.

Matthes, H.W., Maldanado, R., Simonin, F., et al. 1996. Loss of morphine-induced analgesia, reward effect and withdrawal symptoms in mice lacking the μ-opioid-receptor gene. *Nature* 383:819–23.

Matthes, H.W., Smadja, C., Velverde, O., et al. 1998. Activity of the delta-opioid receptor is partially reduced, whereas activity of the κ-receptor is maintained in mice lacking the μ-receptor. *J Neurosci* 18:7285–95.

Matthews, T., Grigore, M., Tang, L., Doat, M., Kow, L., and Pfaff, D. 1997. Sexual reinforcement in the female rat. *J Exp Analysis Behav* 68:399–410.

Matthews, T., Rocca, A., Filipink, R., Staubli, U., Kow, L.-M., and Pfaff, D. 1999.

Approach behavior between male and female rats can reflect affiliative as well as sexual motivation. *Soc Neurosci Abstr* 29:73.

Mattia, A., Farmer, S.C., Takemori, A.E., Sultana, M., Portoghese, P.S., Mosberg, H.I., Bowen, W.D., and Porreca, F. 1992. Spinal opioid δ antinociception in the mouse: mediation by a 5′-NTII-sensitive δ receptor subtype. *J Pharmacol Exp Ther* 260:518–25.

Mayer, D.J., and Hayes, R.L. 1975. Stimulation-produced analgesia: development of tolerance and cross-tolerance to morphine. *Science* 188:941–43.

Mayer, D.J., and Liebeskind, J.C. 1974. Pain reduction by focal electrical stimulation of the brain: an anatomical and behavioral analysis. *Brain Res* 68:73–94.

Mayer, D.J., and Price, D.D. 1976. Central nervous system mechanisms of analgesia. *Pain* 2:379–404.

Mayer, D.J., Wolfle, T.L., Akil, H., Carder, B., and Liebeskind, J.C. 1971. Analgesia from electrical stimulation in the brainstem of the rat. *Science* 174:1351–54.

McCarthy, M.M., Malik, K.F., and Feder, H.H. 1990. Increased GABAergic transmission in medial hypothalamus facilitates lordosis but has the opposite effect in preoptic area. *Brain Res* 507(1):40–44.

McCarthy, M.M., Masters, D.B., Fiber, J.M., Lopez-Colome, A.M., Beyer, C., Komisaruk, B.R., and Feder, H.H. 1991. GABAergic control of receptivity in the female rat. *Neuroendocrinology* 53(5):473–79.

McCarthy, M., Schlenker, E., and Pfaff, D. 1993. Enduring consequences of neonatal treatment with antisense oligonucleotides to estrogen receptor mRNA on sexual differentiation of rat brain. *Endocrinology* 133:433–43.

McGivern, R.F., and Berntson, G.G. 1980. Mediation of diurnal fluctuations in pain sensitivity in the rat by food-intake patterns: reversal by naloxone. *Science* 210:210–11.

McGowan, M.K., and Hammond, D.L. 1993a. Antinociception produced by microinjection of l-glutamate into the ventromedial medulla of the rat: mediation by spinal GABA-A receptors. *Brain Res* 620:86–96.

———. 1993b. Intrathecal GABA-B antagonists attenuate the antinociception produced by microinjection of L-glutamate into the ventromedial medulla of the rat. *Brain Res* 607:39–46.

McKenzie, F., Hunter, A., Daly, C., and Wilson, C. 1984. Evidence that the dopaminergic incerto-hypothalamic tract has a stimulatory effect on ovulation and gonadotrophin release. *Neuroendocrinology* 39:289–95.

McQueen, J., Wilson, H., Sumner, B., and Fink, G. 1999. Serotonin transporter (SERT) mRNA and binding site densities in male rat brain affected by sex steroids. *Mol Brain Res* 63:241–47.

Medina, V., Gupta, D., and Gintzler, A. 1995. Spinal cord dynorphin precursor intermediates decline during pregnancy. *J Neurochem* 65:1374–80.

Medina, V.M., Dawson-Basoa, M.B., and Gintzler, A.R. 1993. 17-β-estradiol and progesterone positively modulate spinal cord dynorphin: relevance to the analgesia of pregnancy. *Neuroendocrinology* 58:310–15.

Melis, M., and Argiolas, A. 1995. Dopamine and sexual behavior. *Neurosci Biobehav Rev* 19:19–38.

Mendell, L. 1966. Physiological properties of unmyelinated fiber projection to the spinal cord. *Exp Neurol* 16:316–32.

Mendelsohn, M., and Karas, R. 1999. The protective effects of estrogen on the cardiovascular system. *N Engl J Med* 340:1801–11.

Mermelstein, P., Becker, J., and Surmeier, D. 1996. Estrogen reduces calcium currents in rat neostriatal neurons via a membrane receptor. *J Neurosci* 16:595–604.

Merskey, H., and Bogduk, N.E., eds. 1994. *Classification of Chronic Pain: Descriptions of Chronic Pain Syndromes and Definitions of Pain Terms.* Seattle: IASP.

Messing, R.B., Vasquez, B.J., Spiehler, V.R., Martinez, J.L., Jensen, R.A., Rigter, H., and McGaugh, J. 1980. $^3$H-dihydromorphine binding in the brain regions of young and aged rats. *Life Sci* 26:921–27.

Mesulam, M. 1995. Cholinergic pathways and the ascending reticular activating system of the human brain. *Ann NY Acad Sci* 757:169–79.

Meunier, J.C., Mollereau, C., Toll, L., et al. 1995. Isolation and structure of the endogenous agonist of the opioid receptor like ORL1 receptor. *Nature* 377: 532–35.

Meyerson, B.J., and Lindstrom, L. 1971. Sexual motivation in the estrogen treated ovariectomized rat. In *Hormonal Steroids*, ed. V.H.T. James and L. Martini, 731–37. Amsterdam: Excerpta Medica.

Meyerson, B.J., and Lindstrom, L.H. 1973. Sexual motivation in the female rat: a methodological study applied to the investigation of the effect of estradiol benzoate. *Acta Physiologica Scandinavica* 389:1–80.

Miczek, K.A., Thompson, M.L., and Schuster, L. 1982. Opioid-like analgesia in defeated mice. *Science* 215:1520–22.

———. 1985. Naloxone injections into periaqueductal gray area and arcuate nucleus block analgesia in defeated mice. *Psychopharmacology* 87:39–42.

Miczek, K., Thompson, M.L., and Schuster, L. 1986. Analgesia following defeat in an aggressive encounter: development of tolerance and changes in opioid receptors. *Ann NY Acad Sci* 467:14.

Millan, M.J., Czlonowski, A., Morris, B., Stein, C., Arendt, R., Huber, A., Hollt, V., and Herz, A. 1988. Inflammation of the hind limb as a model of unilateral localized pain: influence on multiple opioid systems in the spinal cord of the rat. *Pain* 35:299–312.

Millan, M.J., Czlonowski, A., Pilcher, C.W.T., Almeida, O.F.X., Millan, M.H., Colpaert, F.C., and Herz, A. 1987. A model of chronic pain in the rat: functional correlates of alterations in the activity of opioid systems. *J Neurosci* 7:77–87.

Millan, M.J., Gramsch, C., Przewlocki, R., Hollt, V., and Herz, A. 1980. Lesions of the hypothalamic arcuate nucleus produce a temporary hyperalgesia and attenuate the analgesia evoked by stress. *Life Sci* 27:1513–23.

Millan, M.J., Millan, M.H., Czlonkowski, A., Hollt, V., Pilcher, C.W.T., Herz, A., and Colpaert, F.C. 1986. A model of chronic pain in the rat: response of multiple opioid systems to adjuvant-induced arthritis. *J Neurosci* 6:899–906.

Millan, M.J., Przewlocki, R., and Herz, A. 1980. A non-β-endorphinergic adeno-hypophysial mechanism is essential for an analgesic response to stress. *Pain* 8:343–53.

Millan, M.J., Schmauss, C., Millan, M.H., and Herz, A. 1984. Vasopressin and oxy-tocin in the rat spinal cord: analysis of their role in control of nociception. *Brain Res* 309:384–88.

Miller, N.E. 1959. Liberalization of basic S-R concepts: extensions to conflict be-havior, motivation and social learning. In *Psychology: A Study of a Science*, ed. S. Koch, vol. 1, Study 1. New York: McGraw Hill.

———. 1965. Chemical coding of behavior in the brain. *Science* 148:328–38.

Miller, R., ed. 1994. *Anesthesia*. New York: Churchill-Livingstone.

Mitani, A., Ito, K., Hallanger, A.E., Wainer, B.H., Kataoka, K., and McCarley, R.W. 1988. Cholinergic projections from the laterodorsal and pedunculopontine tegmental nuclei to the gigantocellular tegmental field in the cat. *Brain Res* 451:397–402.

Mitchell, J.M., Lowe, D., and Fields, H.L. 1998. The contribution of the rostral ventromedial medulla to the antinociceptive effects of systemic morphine in restrained and unrestrained rats. *Neuroscience* 87:123–33.

Miyamoto, Y., Morita, N., Kitabata, Y., Yamanishi, T., Kishioka, S., Ozaki, M., and Yamamoto, H. 1991. Antinociceptive synergism between supraspinal and spinal sites after subcutaneous morphine evidenced by CNS morphine con-tent. *Brain Res* 552:136–40.

Mobbs, C.C., Harlan, R.R., Burrous, M.M., and Pfaff, D.W. 1988. An estradiol-induced potein synthesized in the ventral medial hypothalamus and trans-ported to the midbrain central gray. *J Neurosci* 8:113–18.

Mogil, J. 1999. The genetic mediation of individual differences in sensitivity to pain and its inhibition. *Proc Natl Acad Sci USA* 96:7744–51.

Mogil, J., and Basbaum, A. 2000. Pain genes?: natural variation and transgenic mutants. *Annu Rev Neurosci* 23:777–811.

Mogil, J., Chesler, E., Wilson, S., Juraska, J.M., and Sternberg, W. 2000. Sex dif-ferences in thermal nociception and morphine antinociception in rodents depend on genotype. *Neurosci Biobehav Rev* 24:375–89.

Mogil, J.S., and Belknap, J.K. 1997. Sex and genotype determine the selective ac-tivation of neurochemically-distinct mechanisms of swim stress-induced anal-gesia. *Pharmacol Biochem Behav* 56:61–66.

Mogil, J.S., Grisel, J.E., Reinscheid, K.K., Civelli, O., Belknap, J.K., and Grandy, D.K. 1996a. Orphanin FQ is a functional anti-opioid peptide. *Neuroscience* 75:333–37.

Mogil, J.S., Grisel, J.E., Zhangs, G., Belknap, J.K., and Grandy, D.K. 1996b. Functional antagonism of μ, δ and κ-opioid antinociception by orphanin FQ. *Neurosci Lett* 214:131–34.

Mogil, J.S., Kest, B., Sadowski, B., and Belknap, J.K. 1996c. Differential genetic mediation of sensitivity to morphine in genetic models of opiate antinociception: influence of nociceptive assay. *J Pharmacol Exp Ther* 276:532–44.

Mogil, J.S., Lichtensteiger, C.A., and Wilson, S.G. 1998. The effect of genotype on sensitivity to inflammatory nociception: characterization of resisitant (A/J) and sensitive (C57BL/6J) inbred mouse strains. *Pain* 76:115–25.

Mogil, J.S., Richards, S.P., O'Toole, L.A., Helms, M.L., Mitchell, S.R., and Belknap, J.K. 1997a. Genetic sensitivity to hot-plate nociception in DBA/2J and C57BL/6J inbred mouse strains: possible sex-specific mediation by δ-2 opioid receptors. *Pain* 70:267–77.

Mogil, J.S., Richards, S.P., O'Toole, L.A., Helms, M.L., Mitchell, S.R., Kest, B., and Belknap, J.K. 1997b. Identification of a sex-specific quantitative trait locus mediating nonopioid stress-induced analgesia in female mice. *J Neurosci* 17:7995–8002.

Mogil, J.S., Sternberg, W.F., Balian, H., Liebeskind, J.C., and Sadowski, B. 1996d. Opioid and non-opioid swim stress-induced analgesia: a parametric analysis. *Physiol Behav* 59:123–32.

Mogil, J.S., Sternberg, W.F., Kest, B., Marek, P., and Liebeskind, J.C. 1993. Sex differences in the antagonism of swim stress-induced analgesia: effects of gonadectomy and estrogen replacement. *Pain* 53:17–25.

Mogil, J.S., Wilson, S.G., Bon, K., et al. 1999a. Heritability of nociception I: responses of 11 inbred mouse strains on 12 measures of nociception. *Pain* 80:-67–82.

———. 1999b. Heritability of nociception II: types of nociception revealed by genetic correlation analysis. *Pain* 80:83–93.

Mollereau, C., Parmentier, M., Mailleux, P., Butour, J.L., Moisand, C., Chalon, P., Caput, D., Vassart, G., and Meunier, J.C. 1994. ORL-1, a novel member of the opioid family: cloning, functional expression and localization. *FEBS Lett* 341:33–38.

Monroe, P.J., Hawranko, A.A., Smith, D.L., and Smith, D.J. 1996. Biochemical and pharmacological characterization of multiple β-endorphinergic antinociceptive systems in the rat periaqueductal gray. *J Pharmacol Exp Ther* 276:65–93.

Monroe, P.J., Smith, D.L., and Smith, D.J. 1997. β-endorphin dose-dependently activates multiple spinopetal antinociceptive pathways following its microinjection into the rat periaqueductal gray: comparison with those activated by morphine. *Analgesia* 3:21–26.

Moore, R., Halaris, A., and Jones, B. 1978. Serotonin neurons of the midbrain raphe: ascending projections. *J Comp Neurol* 180:417–38.

Moreau, J.L., and Fields, H.L. 1986. Evidence for GABA involvement in midbrain control of medullary neurons that modulate nociceptive transmission. *Brain Res* 397(1):37–46.

Morgan, C.T. 1943. *Physiological Psychology.* New York: McGraw Hill.

Morgan, M., Heinricher, M., and Fields, H. 1992. Circuitry linking opioid-sensitive nociceptive modulatory systems in periaqueductal gray and spinal cord with rostral ventromedial medulla. *Neuroscience* 47:863–71.

Morgan, M., Sohn, J.-H., and Liebeskind, J. 1989. Stimulation of the periaqueductal gray matter inhibits nociception at the supraspinal as well as spinal level. *Brain Res* 502:61–66.

Morilak, D., Fornal, C., and Jacobs, B. 1987. Effects of physiological manipulations on locus coeruleus neuronal activity in freely moving cats. I. Thermoregulatory challenge. *Brain Res* 422:17–23.

Moruzzi, G., and Magoun, H.W. 1949. Brain stem reticular formation and activation of the EEG. *EEG Clin Neurophysiol* 1:455–73.

Moskowitz, A.S., and Goodman, R.R. 1985. Autoradiographic analysis of $\mu$-1 and $\mu$-2 and $\delta$ opioid binding in the central nervous system of C57BL6BY and CXBK (opioid receptor deficient) mice. *Brain Res* 360:108–16.

Moss, R., and McCann, S. 1973. Induction of mating behavior in rats by luteinizing hormone-releasing factor. *Science* 181:177–79.

Moss, R., Qin, G., and Wong, M. 1997. Estrogen: nontranscriptional signaling pathway. In *Recent Progress in Hormone Research,* vol. 52, ed. P. Conn, 33–700. Bethesda, MD: The Endocrine Society.

Mountz, J., Bradley, L., Modell, J., Alexander, R., Triana-Alexander, M., Aaron, L., Stewart, K., Alarcon, G., and Mountz, J. 1995. Fibromyalgia in women: abnormalities of regional cerebral blood flow in the thalamus and the caudate nucleus are associated with low pain thresholds. *Arthritis Rheum* 38: 926–38.

Mousa, S., Miller, C.H., and Couri, D. 1981. Corticosteroid modulation and stress-induced analgesia. *Neuroendocrinology* 33:317–19.

———. 1983. Dexamethasone and stress-induced analgesia. *Psychopharmacology* 79:199–202.

Mouton, L., and Holstege, G. 1994. The periaqueductal gray in the cat projects to Lamina VIII and the medial part of Lamina VII throughout the length of the spinal cord. *Exp Brain Res* 101:253–64.

———. 1998. Three times as many Lamina I neurons project to the periaqueductal gray than to the thalamus: a retrograde study in the cat. *Neurosci Lett* 255:107–10.

———. 2001. The segmental and laminar origin of the spinal neurons projecting to the periaqueductal gray (PAG) in the cat suggests the existence of at least five separate Spino-PAG systems. *J Comp Neurol.* In press.

Mouton, L., Kerstens, L., and Holstege, G. 1996. Dorsal border periaqueductal gray neurons project to the area directly adjacent to the central canal ependyma of the C4-T8 spinal cord. *Exp Brain Res* 112:11–23.

Moye, T.B., Hyson, R.L., Grau, J.W., and Maier, S.F. 1983. Immunization of opi-oid analgesia: effects of prior escapable shock on subsequent shock-induced and morphine-induced antinociception. *Learn Motivat* 14:238–51.

Munk, M., Roelfsema, P.R., Konig, P., Engel, A.K., and Singer, W. 1996. Role of reticular activation in the modulation of intracortical synchronization. *Science* 272:271–74.

Nakahara, D., and Nakamura, M. 1999. Differential effect of immobilization stress on in vivo synthesis rate of monoamines in medial prefrontal cortex and nucleus accumbens of conscious rats. *Synapse* 32:238–42.

Nasstrom, J., Boo, E., Stahlberg, M., and Berge, O.-G. 1993. Tissue distribution of two NMDA receptor antagonists, [3H]CGS19755 and [$^{3}$H]MK-801, after intrathecal injection in mice. *Pharmacol Biochem Behav* 44:9–15.

Nasstrom, J., Karlsson, U., and Berge, O.-G. 1993. Systemic or intracerebroven-tricular injection of NMDA receptor antagonists attenuates the antinocicep-tive activity of intrathecally administered NMDA receptor antagonists. *Brain Res* 623:47–55.

Nichols, D., Thorn, B., and Berntson, G. 1989. Opiate and serotonergic mech-anisms of stimulation-produced analgesia within the periaqueductal gray. *Brain Res Bull* 22:717–24.

Nicot, A., Ogawa, S., Berman, Y., Carr, K., and Pfaff, D. 1997. Effects of an intra-hypothalamic injection of antisense oligonucleotides for preproenkephalin mRNA in female rats: evidence for opioid involvement in lordosis reflex. *Brain Res* 777:60–68.

Nimmo, W., ed. 1994. *Anesthesia*. Oxford: Blackwell.

Nishizuka, M., and Pfaff, D.W. 1989. Intrinsic synapses in the ventromedial nu-cleus of the hypothalamus: an ultrastructural study. *J Comp Neurol* 286(2): 260–68.

Nissen, H.W. 1929. Experiments on sex drive in rats. *Genet Psychol Monogr* 5:451–548.

Nonaka, R., and Moroji, T. 1984. Quantitative autoradiography of muscarinic cholinergic receptors in the rat brain. *Brain Res* 296:295–303.

Nuseir, K., and Proudfit, H. 2000. Bidirectional modulation of nociception by GABA neurons in the dorsolateral pontine tegmentum that tonically inhibit spinally projecting noradrenergic A-7 neurons. *Neuroscience* 96:773–83.

Nuseir, K., Heidenreich, B.A., and Proudfit, H.K. 1999. The antinociception pro-duced by microinjection of a cholinergic agonist in the ventromedial me-dulla is mediated by noradrenergic neurons in the A7 catecholamine cell group. *Brain Res* 822:1–7.

O'Connor, L.H., Nock, B., and McEwen, B.S. 1988. Regional specificity of γ-amino-butyric acid receptor regulation by estradiol. *Neuroendocrinology* 47(6):473–81.

O'Connor, P., and Chipkin, R.E. 1984. Comparisons between warm and cold wa-ter swim stress in mice. *Life Sci* 35:631–39.

Ogawa, S., Eng, V., Taylor, J., Lubahn, D., Korach, K., and Pfaff, D. 1998a. Roles of estrogen receptor-α gene expression in reproduction-related behaviors in female mice. *Endocrinology* 139:5070–81.

Ogawa, S., Gordan, J., Taylor, J., Lubahn, D., Korach, K., and Pfaff, D. 1996a. Reproductive functions illustrating direct and indirect effects of genes on behavior. *Horm Behav* 30:487–94.

Ogawa, S., Gustafsson, J., Korach, K., and Pfaff, D. 1999. Survival of reproduction-related behaviors in male and female estrogen receptor b deficient (bERKO) male and female mice. *Proc Natl Acad Sci USA* 96:12887–92.

Ogawa, S., Kow, L.M., and Pfaff, D.W. 1994. In vitro electrophysiological characterization of midbrain periaqueductal gray neurons in female rats: responses to GABA-and met-enkephalin-related agents. *Brain Res* 666:239–49.

Ogawa, S., Lubahn, D.B., Korach, K.S., and Pfaff, D.W. 1997. Behavioral effects of estrogen receptor gene disruption in male mice. *Proc Natl Acad Sci USA* 94:1476–81.

Ogawa, S., Taylor, J., Lubahn, D., Korach, K., and Pfaff, D. 1996b. Reversal of sex roles in genetic female mice by disruption of estrogen receptor gene. *Neuroendocrinology* 64:467–70.

Ogawa, S., Washburn, T., Taylor, J., Lubahn, D., Korach, K., and Pfaff, D. 1998b. Modifications of testosterone-dependent behaviors by estrogen receptor-alpha gene disruption in male mice. *Endocrinology* 139:5058–69.

Oka, T., Oka, K., Hosoi, M., Aou, S., and Ho, T. 1995. The opposing effects of interleukin-1 b microinjected into the preoptic hypothalamus and the ventromedial hypothalamus on nociceptive behavior in rats. *Brain Res* 700:271–78.

Okuda-Ashitake, E., Minami, T., Tachibana, S., Yoshihara, Y., Nishiuchi, Y., Kimura, T., and Ito, S. 1998. Nocistatin, a peptide that blocks nociceptin action in pain transmission. *Nature* 392:286–89.

Oliveras, J., and Besson, J. 1988. Stimulation-produced analgesia in animals: behavioural investigations. *Prog Brain Res* 77:141–57.

Oliveras, J.L., Besson, J.M., Guilbaud, G., and Liebeskind, J.C. 1974. Behavioral and electrophysiological evidence of pain inhibition from midbrain stimulation in the cat. *Exp Brain Res* 20:32–44.

Oliveras, J.L., Hosobuchi, Y., Guilbaud, G., and Besson, J.M. 1978. Analgesic electrical stimulation of the feline nucleus raphe magnus: development of tolerance and its reversal by 5-HTP. *Brain Res* 146:404–9.

Osaka, T., and Matsumura, H. 1994. Noradrenergic inputs to sleep-related neurons in the preoptic area from the locus coeruleus and the ventrolateral medulla in the rat. *Neurosci Res* 19:39–50.

Palkovits, M., Baffi, J., and Pacak, K. 1999. The role of ascending neuronal pathways in stress-induced release of noradrenaline in the hypothalamic paraventricular nucleus of rats. *J Neuroendocrinol* 11:529–39.

Pan, Y.X., Cheng, J., and Pasternak, G.W. 1994. Isolation and characterization of a κ3 related opioid receptor. *Regul Pept* 54:217–18.

Pan, Y.X., Cheng, J., Xu, J., Rossi, G., Jacobson, E., Ryan-Moro, J., Brooks, A.I., Dean, G.E., Standifer, K.M., and Pasternak, G.W. 1995. Cloning and functional characterization through antisense mapping of a κ3-related opioid receptor. *Mol Pharmacol* 47:1180–88.

Pan, Z., Hirakawa, N., and Fields, H. 2000. A cellular mechanism for the bidirectional pain-modulating actions of orphanin FQ/nociceptin. *Neuron* 26: 515–22.

Pan, Z., Tershner, S., and Fields, H. 1997. Cellular mechanism for anti-analgesic action of agonists of the κ-opioid receptor. *Nature* 389:382–85.

Pan, Z., Williams, J., and Osborne, P. 1990. Opioid actions on single nucleus raphe magnus neurons from rat and guinea pig in vitro. *J Physiol* 427:519–32.

Pan, Z.Z., and Fields, H.L. 1996. Endogenous opioid-mediated inhibition of putative pain-modulating neurons in rat rostral ventromedial medulla. *Neuroscience* 74:855–62.

Papka, R., Srinivasan, B., Miller, K., and Hayashi, S. 1997. Localization of estrogen receptor protein and estrogen receptor messenger RNA in peripheral autonomic and sensory neurons. *Neuroscience* 79:1153–63.

Pare, W.P. 1969. Age, sex, and strain differences in the aversive threshold to grid shock in the rat. *J Comp Physiol Psychol* 69:214–18.

Parry, B. 1989. Reproductive factors affecting the course of affective illness in women. *Women's Dis* 12:207–20.

Pasternak, G.W. 1980. Multiple opiate receptors: [$^3$H]ethylketocyclazocine receptor binding and ketocyclazocine analgesia. *Proc Natl Acad Sci USA* 77: 3691–94.

Pasternak, G.W., and Standifer, K.M. 1995. Mapping of opioid receptors using antisense oligodeoxynucleotides: correlating their molecular biology and pharmacology. *Trends Pharmacol Sci* 16:344–50.

Pasternak, G.W., Bodnar, R.J., Clark, J.A., and Inturrisi, C.E. 1987. Morphine-6-glucuronide, a potent μ agonist. *Life Sci* 41:2845–49.

Pasternak, G.W., Childers, S.R., and Snyder, S.H. 1980. Naloxonazone, a long-acting opiate antagonist: effects in intact animals and on opiate receptor binding in vitro. *J Pharmacol Exp Ther* 214:1311–15.

Pasternak, G.W., and Wood, P.L. 1986. Multiple μ opiate receptors. *Life Sci* 38: 1889–96.

Pasternak, K.R., Rossi, G.C., Zuckerman, A., and Pasternak, G.W. 1999. Antisense mapping KOR-1: evidence for multiple κ analgesic mechanisms. *Brain Res* 826:289–92.

Paul, D., and Pasternak, G. 1988. Differential blockade by naloxonazine of two μ opiate actions: analgesia and inhibition of gastrointestinal transit. *Eur J Pharmacol* 149:403–4.

Paul, D., and Phillips, A.G. 1986. Selective effects of pirenpirone on analgesia produced by morphine or electrical stimulation at sites in the nucleus raphe magnus and periaqueductal gray. *Psychopharmacology* 88:172–76.

Paul, D., Bodnar, R.J., Gistrak, M.A., and Pasternak, G.W. 1989. Different μ receptor subtypes mediate spinal and supraspinal analgesia in mice. *Eur J Pharmacol* 168:307–14.

Paul, D., Levison, J.A., Howard, D.H., Pick, C.G., Hahn, E.F., and Pasternak, G.W. 1990. Naloxone benzoylhydrazone (NalBxoH) analgesia. *J Pharmacol Exp Ther* 255:769–74.

Pavlovic, Z., and Bodnar, R. 1998a. Opioid supraspinal analgesic synergy between the amygdala and periaqueductal gray in rats. *Brain Res* 779:158–69.

Pavlovic, Z., and Bodnar, R.J. 1993. Antinociceptive and hypothermic cross-tolerance between continuous and intermittent cold-water swims in rats. *Physiol Behav* 54:1081–84.

Pavlovic, Z., Cooper, M.L., and Bodnar, R. 1996a. Enhancements in swim stress-induced hypothermia, but not analgesia following amygdala lesions in rats. *Physiol Behav* 59:77–82.

Pavlovic, Z., Cooper, M.L., and Bodnar, R.J. 1996b. Opioid antagonists in the periaqueductal gray inhibit morphine and β-endorphin analgesia elicited from the amygdala of rats. *Brain Res* 741:13–26.

Pavlovic, Z.W., and Bodnar, R.J. 1998b. U50488H-induced analgesia in the amygdala: test-specific effects and blockade by opioid antagonists in the periaqueductal gray. *Analgesia* 3:223–30.

Pazos, A., Cortes, R., and Palacios, J.M. 1985. Quantitative autoradiographic mapping of serotonin receptors in the rat brain. II. Serotonin-2 receptors. *Brain Res* 346:231–49.

Pecins-Thompson, M., and Bethea, C. 1998. Ovarian steroid regulation of serotonin-1A autoreceptor messenger RNA expression in the dorsal raphe of rhesus macaques. *Neuroscience* 89:267–77.

Pecins-Thompson, M., Brown, N., and Bethea, C. 1998. Regulation of serotonin re-uptake transporter mRNA expression by ovarian steroids in rhesus macaques. *Brain Res Mol Brain Res* 53:120–29.

Peirce, J.T., and Nuttall, R.L. 1961. Self-paced sexual behavior in the female rat. *J Comp Physiol Psychol* 54:310–13.

Pert, C.B., and Snyder, S.H. 1973. Opiate receptor: demonstration in nervous tissue. *Science* 179:1011–14.

Petersen, S.L., and LaFlamme, K.D. 1997. Progesterone increases levels of μ-opioid receptor mRNA in the preoptic area and arcuate nucleus of ovariectomized, estradiol-treated female rats. *Mol Brain Res* 52:32–37.

Pfaff, D. 1973. Luteinizing hormone releasing factor (LRF) potentiates lordosis behavior in hypophysectomized ovariectomized female rats. *Science* 182:1148–49.

———. 1982. Neurobiological mechanisms of sexual motivation. In *The Physiological Mechanisms of Motivation,* ed. D.W. Pfaff, 287–317. New York: Springer-Verlag.

Pfaff, D., and Lewis, C. 1974. Film analyses of lordosis in female rats. *Horm Behav* 5:317–35.

Pfaff, D.W. 1980. *Estrogens and Brain Function: Neural Analysis of a Hormone Controlled Mammalian Reproductive Behavior.* New York: Springer-Verlag.

———. 1999a. *Drive: Neural and Molecular Mechanisms for Sexual Motivation.* Cambridge, Mass.: MIT Press.

———. 1999b. Introduction: genetic influences on the nervous system and behavior. In *Genetic Influences on Neural and Behavioral Function*, eds D.W. Pfaff, W. Berrettini, T. Joh, and S. Maxson. Boca Raton: CRC Press.

Pfaff, D.W., and Keiner, M. 1973. Atlas of estradiol-concentrating cells in the central nervous system of the female rat. *J Comp Neurol* 151:121–58.

Pfaff, D.W., and Sakuma, Y. 1979a. Deficit in the lordosis reflex of female rats caused by lesions in the ventromedial nucleus of the hypothalamus. *J Physiol (Lond)* 288:203–10.

———. 1979b. Facilitation of the lordosis reflex of female rats from the ventromedial nucleus of the hypothalamus. *J Physiol (Lond)* 288:189–202.

Pfaff, D.W., and Schwartz-Giblin, S. 1988. Cellular mechanisms of female reproductive behaviors. In *The Physiology of Reproduction,* ed. E. Knobil and J. Neill, 1487–1569. New York: Raven Press.

Pfaff, D.W., Kow, L.M., Zhu, Y.S., Scott, R.E.M., Wu-Peng, S.X., and Dellovade, T. 1996. Hypothalamic cellular and molecular mechanisms helping to satisfy axiomatic requirements for reproduction. *J Neuroendocrinol* 8:325–36.

Pfaus, J., and Pfaff, D. 1992. m-, d-and k-opioid receptor agonists selectively modulate sexual behaviors in the female rat: differential dependence on progesterone. *Horm Behav* 26:457–73.

Pfaus, J., Smith, W., and Coopersmith, C. 1999. Appetitive and consummatory sexual behaviors of female rats in bilevel chambers. I. A correlational and factor analysis and the effects of ovarian hormones. *Horm Behav* 35:224–40.

Phoenix, C.H., Goy, R.W., Gerall, A.A., and Young, W.C. 1959. Organizing action of prenatally administered testosterone propionate on the tissues mediating mating behavior in the female guinea pig. *Endocrinology* 65:369–82.

Pick, C.G., Nejat, R.J., and Pasternak, G.W. 1993. Independent expression of two pharmacologically distinct supraspinal μ analgesic systems in genetically-different mouse strains. *J Pharmacol Exp Ther* 265:166–71.

Pick, C.G., Roques, B., Gacel, G., and Pasternak, G.W. 1992. Supraspinal μ-2 receptors mediate spinal/supraspinal morphine synergy. *Eur J Pharmacol* 220: 275–77.

Pleim, E., Lisciotto, C., and DeBold, J. 1990. Facilitation of sexual receptivity in hamsters by simultaneous progesterone implants in the VMH and the ventral mesencephalon. *Horm Behav* 24:139–51.

Plum, F. 1991. Coma and related global disturbances of the human conscious state. In *Cerebral Cortex,* vol. 9, ed. A. Peters, 359–425. New York: Plenum Publishing Corporation.

Plum, F., and Posner, J.B. 1982. *The Diagnosis of Stupor and Coma*, 3rd ed. Philadelphia: Davis.

Plum, F., Schiff, N., Ribary, U., and Llinas, R. 1998. Coordinated expression in chronically unconscious persons. *Phil Trans R Soc Lond* 353:1929–33.

Porreca, F., Mosberg, H.I., Hurst, R., Hruby, V.J., and Burks, T.F. 1984. Roles of μ, δ and κ opioid receptors in spinal and supraspinal mediation of gastrointestinal transit effects and hot-plate analgesia in the mouse. *J Pharmacol Exp Ther* 230:341–48.

Posner, M. 1995. Attention in cognitive neuroscience: an overview. In *The Cognitive Neurosciences*, M. Gazzaniga, 615–24. Cambridge, Mass.: MIT Press.

Posner, M., and Petersen, S. 1990. The attention system of the human brain. *Annu Rev Neurosci* 13:25–42.

Potrebic, S.B., Fields, H.L., and Mason, P. 1994. Serotonin immunoreactivity is contained in one physiological cell class in the rat rostral ventromedial medulla. *J Neurosci* 14:1655–65.

Potrebic, S.B., Mason, P., and Fields, H.L. 1995. The density and distribution of serotonergic appositions onto identified neurons in the rat rostral ventromedial medulla. *J Neurosci* 15:3273–83.

Powley, T.L. 1977. The ventromedial hypothalamic syndrome, satiety and a cphalic phase hypothesis. *Psychol Rev* 84:89–126.

Pott, C., Kramer, S., and Siegel, A. 1987. Central gray modulation of affective defense is differentially sensitive to naloxone. *Physiol Behav* 40:207–13.

Priest, C.A., Eckersell, C.B., and Micevych, P.E. 1995. Estrogen regulates preproenkephalin-A mRNA levels in the rat ventromedial nucleus: temporal and cellular aspects. *Mol Brain Res* 28:251–62.

Priest, C., Borsook, D., and Pfaff, D. 1996. Estrogen and stress interact to regulate the hypothalamic expression of a human proenkephalin promoter-(-galactosidase fusion) gene in a site-specific and sex-specific manner. *J Neuroendocrinol* 9:317–26.

Prieto, G., Cannon, J., and Liebeskind, J. 1983. Nucleus raphe magnus lesions disrupt stimulation produced analgesia from ventral but not dorsal midbrain areas in the rat. *Brain Res* 261:53–57.

Proudfit, H., and Yeomans, D. 1995. The modulation of nociception by enkephalin-containing neurons in the brainstem. In *The Pharmacology of Opioid Peptides*, ed. L. Tseng, 197–217. Amsterdam: Harwood Academic.

Proudfit, H.K. 1980. Reversible inactivation of raphe magnus neurons: effects on nociceptive threshold and morphine-induced analgesia. *Brain Res* 201:459–64.

———. 1988. Pharmacologic evidence for the modulation of nociception by noradrenergic neurons. *Prog Brain Res* 88:359–72.

———. 1992. The behavioral pharmacology of the noradrenergic descending system. In *Towards the Use of Noradrenergic Agonists for the Treatment of Pain*. J.M. Besson, 119–36. Amsterdam: Elsevier.

Proudfit, H.K., and Clark, F.M. 1991. The projections of locus coeruleus neurons to the spinal cord. *Prog Brain Res* 88:123–41.

Proudfit, H.K., and Monsen, M. 1999. Ultrastructural evidence that substance P neurons form synapses with noradrenergic neurons in the A7 catecholamine cell group that modulate nociception. *Neuroscience* 91:1499–1513.

Przewlocki, R., Shearman, G.T., and Herz, A. 1983. Mixed opioid/nonopioid effects of dynorphin and dynorphin-related peptides after their intrathecal injection in rats. *Neuropeptides* 3:233–40.

Quiñones-Jenab, V., Ogawa, S., Jenab, S., and Pfaff, D.W. 1996. Estrogen regulation of preproenkephalin messenger RNA in the forebrain of female mice. *J Chem Neuroanat* 12:29–36.

Quiñones-Jenab, V., Jenab, S., Ogawa, S., Inturrisi, C., and Pfaff, D.W. 1997. Estrogen regulation of mu-opioid receptor messenger RNA in the forebrain of female rats. *Mol Brain Res* 47:134–138.

Quirion, R., and Weiss, A.S. 1983. Peptide E and other proenkephalin-derived peptides are potent κ opiate receptor agonists. *Peptides* 4:445–49.

Raab, H., Pilgrim, C., and Reisert, I. 1995. Effects of sex and estrogen on tyrosine hydroxylase mRNA in cultured embryonic rat mesencephalon. *Mol Brain Res* 33:157–64.

Rachman, I., Unnerstall, J., Pfaff, D., and Cohen, R. 1998. Estrogen alters behavior and forebrain c-fos expression in ovariectomized rats subjected to the forced swim test. *Proc Natl Acad Sci USA* 95:13941–46.

Ragnauth, A., Schuller, A., Chan, J., Ogawa, S., Bodnar, R., Pintar, J., and Pfaff, D. 2001. Female preproenkephalin knockout mice display altered emotional responses. *Proc Natl Acad Sci USA* In press.

Ramboz, S., Oosting, R., Amara, D., Kung, H., Blier, P., Mendelsohn, M., Mann, J., Brunner, D., and Hen, R. 1998. Serotonin receptor 1A knockout: an animal model of anxiety-related disorder. *Proc Natl Acad Sci USA* 95:14476–81.

Ranaldi, R., Pocock, D., Zereik, R., and Wise, R. 1999. Dopamine fluctuations in the nucleus acumbens during maintenance, extinction, and reinstatement of intravenous D-amphetamine self-administration. *J Neurosci* 19:4102–9.

Rasmussen, K., and Jacobs, B. 1986. Single unit activity of locus coeruleus neurons in the freely moving cat. II. Conditioning and pharmacologic studies. *Brain Res* 371:335–44.

Rasmussen, K., Morilak, D., and Jacobs, B. 1986. Single unit activity of locus coeruleus neurons in the freely moving cat. I. During naturalistic behaviors and in response to simple and complex stimuli. *Brain Res* 371:324–34.

Reinscheid, R.K., Nothacker, H.P., Bourson, A., Ardati, A., Henningsen, R.A., Bunzow, J.R., Grandy, D.K., Langen, H., Monsma, F.J., and Civilli, O. 1995. Orphanin FQ: a neuropeptide that activates an opioidlike G protein-coupled receptor. *Science* 270:792–94.

Reisert, I., Han, V., Lieth, E., Toran-Allerand, D., Pilgrim, C., and Lauder, J.

1987. Sex steroids promote neurite growth in mesencephalic tyrosine hydroxylase immunoreactive neurons in vitro. *J Dev Neurosci* 5:91–98.

Reisine, T., and Bell, G.I. 1993. Molecular biology of opioid receptors. *Trends Neurosci* 16:506–10.

Renno, W., Mahmoud, M., Hamdi, A., and Beitz, A. 1999. Quantitative immunoelectron microscopic colocalization of GABA and enkephalin in the ventrocaudal periaqueductal gray of the rat. *Synapse* 31:216–28.

Reynolds, D.V. 1969. Surgery in the rat during electrical analgesia induced by focal brain stimulation. *Science* 164:444–45.

Rhodes, D.L. 1979. Periventricular system lesions and stimulation-produced analgesia. *Pain* 7:51–63.

Rhodes, D.L., and Liebeskind, J.C. 1978. Analgesia from rostral brain stem stimulation in the rat. *Brain Res* 143:521–32.

Riedl, M., Shuster, S., Vulchanova, L., Wang, J., Loh, H.H., and Elde, R. 1996. Orphanin FQ/nociceptin-immunoreactive nerve fibers parallel those containing endogenous opioids in rat spinal cord. *Neuroreport* 7:1369–72.

Riley, J., Robinson, M., Wise, E., Myers, C., and Fillingim, R. 1998. Sex differences in the perception of noxious experimental stimuli: a meta-analysis. *Pain* 74:181–87.

Riskind, P., and Moss, R.L. 1983. Effects of lesions of putative LHRH-containing pathways and midbrain nuclei on lordotic behavior and leutinizing hormone release in ovariectomized rats. *Brain Res Bull* 11:493–500.

Rizvi, T.A., Ennis, M., Behbehani, M.M., and Shipley, M.T. 1991. Connections between the central nucleus of the amygdala and the midbrain periaqueductal gray: topography and reciprocity. *J Comp Neurol* 303:121–31.

Robbins, T., and Everitt, B. 1996. Arousal systems and attention. In *Handbook of Cognitive Neuroscience*, ed. M. Gazzaniga, 703–20. Cambridge, Mass.: MIT Press.

Robbins, T., Granon, S., Muir, J., Durantou, F., Harrison, A., and Everitt, B. 1998. Neural systems underlying arousal and attention: implications for drug abuse. *Ann NY Acad Sci* 846:222–37.

Roberts, J.L., Seeburg, P.H., Shine, J., Herbert, E., Baxter, J.D., and Goodman, H.M. 1979. Corticotropin and β-endorphin: construction of analysis of recombinant DNA complementary to mRNA for the common precursor. *Proc Natl Acad Sci USA* 76:2153–q57.

Roberts, L.A., Beyer, C., and Komisaruk, B.R. 1985. Strychnine antagonizes vaginal stimulation-produced analgesia at the spinal cord. *Life Sci* 36:2017.

Robertson, J.A., and Bodnar, R.J. 1993. Site-specific modulation of morphine and swim-induced antinociception following thyrotropin-releasing hormone in the rat periaqueductal gray. *Pain* 55:71–84.

Robertson, J.A., Hough, L.B., and Bodnar, R.J. 1988. Potentiation of opioid and nonopioid forms of swim analgesia by cimetidine. *Pharmacol Biochem Behav* 31:107–12.

Rochford, J., and Dawes, P. 1993. Effect of naloxone on the habituation of novelty-induced hypoalgesia: the collateral inhibition hypothesis revisited. *Pharmacol Biochem Behav* 46:117–13.

Rochford, J., and Henry, J.L. 1988. Analgesia induced by continuous versus intermittent cold-water swim in the rat: differential effects of intrathecal administration of phentolamine and methysergide. *Pharmacol Biochem Behav* 31:27–31.

Roder, S., and Ciriello, J. 1993. Innervation of the amygdaloid complex by catecholaminergic cell groups of the ventrolateral medulla. *J Comp Neurol* 332:105–22.

———. 1994. Collateral axonal projections to limbic structures from ventrolateral medullary A1 noradrenergic neurons. *Brain Res* 638:182–88.

Rodgers, R.J. 1977. Elevation of aversive threshold in rats by intra-amygdaloid injection of morphine sulphate. *Pharmacol Biochem Behav* 6:385–90.

———. 1978. Influence of intra-amygdaloid opiate injections on shock thresholds, tail-flick latencies and open field behaviour in rats. *Brain Res* 153:211–16.

Rodgers, R.J., and Hendrie, C.A. 1983. Social conflict activates status-dependent endogenous analgesic or hyperalgesic mechanisms in male mice: effects of naloxone on nociception and behavior. *Physiol Behav* 30:775–80.

Rodgers, R.J., and Randall, J.I. 1985. Social conflict analgesia: studies on naloxone antagonism and morphine cross-tolerance in male DBA/2 mice. *Pharmacol Biochem Behav* 23:883–87.

———. 1986a. Acute non-opioid analgesia in defeated male mice. *Physiol Behav* 36:947–50.

———. 1986b. Extended attack from a resident conspecific is critical to the development of long-lasting analgesia in male intruder mice. *Physiol Behav* 38:427–30.

———. 1986c. Resident's scent: a critical stimulus in acute analgesic reaction to defeat experience in male mice. *Physiol Behav* 37:317–22.

———. 1987a. Are the analgesic effects of social defeat mediated by benzodiazepine receptors? *Physiol Behav* 41:279–89.

———. 1987b. Benzodiazepine ligands, nociception, and defeat analgesia. *Psychopharmacology* 91:303–7.

———. 1987c. Social conflict analgesia: inhibition of early non-opioid component by diazepam or flumazepil fails to affect appearance of late opioid component. *Brain Res Bull* 19:141.

Roerig, S.C., and Fujimoto, J.M. 1989. Multiplicative interactions between intrathecally and intracerebroventricularly administered μ opioid agonists but limited interactions between δ and κ agonists for antinociception in mice. *J Pharmacol Exp Ther* 249:762–67.

Roerig, S.C., Fujimoto, J.M., and Tseng, L.F. 1988. Comparisons of descending pain inhibitory pathways activated by β-endorphin and morphine as characterized by supraspinal and spinal antinociceptive interactions in mice. *J Pharmacol Exp Ther* 247:1107–13.

Roerig, S.C., O'Brien, S.M., Fujimoto, J.M., and Wilcox, G.L. 1984. Tolerance to morphine analgesia: decreased multiplicative interaction between spinal and supraspinal sites. *Brain Res* 302:360–63.

Romano, G., Harlan, R., Shivers, B., Howells, R., and Pfaff, D. 1988. Estrogen increases proenkephalin messenger ribonucleic acid levels in the ventromedial hypothalamus of the rat. *Mol Endocrinol* 2:1320–28.

Romano, G.J., Krust, A., and Pfaff, D.W. 1989. Expression and estrogen regulation of progesterone receptor mRNA in neurons of the mediobasal hypothalamus: an *in situ* hybridization study. *Mol Endocrinol* 3:1295–1300.

Romano, G.J., Mobbs, C.V., Lauber, A., Howells, R.D., and Pfaff, D.W. 1990. Differential regulation of proenkephalin gene expression by estrogen in the ventromedial hypothalamus of male and female rats: implications for the molecular basis of a sexually differentiated behavior. *Brain Res* 536: 63–68.

Romano, G., Mobbs, C., Howells, R., and Pfaff, D. 1989. Estrogen regulation of proenkephalin gene expression in the ventromedial hypothalamus of the rat: temporal qualities and synergism with progesterone. *Mol Brain Res* 5: 51–58.

Romero, M.-T., and Bodnar, R.J. 1986. Gender differences in two forms of coldwater swim analgesia. *Physiol Behav* 37:893–97.

Romero, M.T., and Bodnar, R.J. 1987. Maintainance of β-endorphin analgesia across age cohorts. *Neurobiol Aging* 8:167–70.

Romero, M.T., Cooper, M.L., Komisaruk, B.R., and Bodnar, R.J. 1988. Gender-specific and gonadectomy-specific effects upon swim analgesia: role of steroid replacement therapy. *Physiol Behav* 44:257–65.

Romero, M.T., Kepler, K.L., and Bodnar, R.J. 1988. Gender determinants of opioid mediation of swim analgesia in rats. *Pharmacol Biochem Behav* 29:705–9.

Romero, M.T., Kepler, K.L., Cooper, M.L., Komisaruk, B.R., and Bodnar, R.J. 1987. Modulation of gender-specific effects upon swim analgesia in gonadectomized rats. *Physiol Behav* 40:39–45.

Rosas-Arellano, M., Solano-Flores, L., and Ciriello, J. 1999. Co-localization of estrogen and angiotensin receptors within subfornical organ neurons. *Brain Res* 837:254–62.

Rose, M.D. 1974. Pain-reducing properties of rewarding electrical brain stimulation in the rat. *J Comp Physiol Psychol* 87:607–17.

Rossi, G.C., Brown, G.P., Leventhal, L., Yang, K., and Pasternak, G.W. 1996. Novel receptor mechanisms for heroin and morphine-6B-glucuronide analgesia. *Neurosci Lett* 216:1–4.

Rossi, G.C., Leventhal, L., and Pasternak, G.W. 1996. Naloxone-sensitive orphanin FQ-induced analgesia in mice. *Eur J Pharmacol* 311:R7–8.

Rossi, G.C., Leventhal, L., Bolan, E., and Pasternak, G.C. 1997a. Pharmacological characterization of orphanin FQ/nociceptin and its fragments. *J Pharmacol Exp Ther* 282:858–65.

Rossi, G.C., Leventhal, L., Pan, Y.-X., Cole, J., Su, W., Bodnar, R.J., and Pasternak, G.W. 1997b. Antisense mapping of MOR-1 in rats: distinguishing between morphine and morphine-6β-glucuronide antinociception. *J Pharmacol Exp Ther* 281:109–14.

Rossi, G.C., Pan, Y.X., Brown, G.P., and Pasternak, G.W. 1995. Antisense mapping the MOR-1 receptor: evidence for alternative splicing and a novel morphine-6β-glucuronide receptor. *FEBS Lett* 369:192–96.

Rossi, G.C., Pasternak, G.W., and Bodnar, R.J. 1993. Synergistic brainstem interactions for morphine analgesia. *Brain Res* 624:171–80.

———. 1994. μ and δ opioid synergy between the periaqueductal gray and the rostro-ventral medulla. *Brain Res* 665:85–93.

Rossi, G.C., Standifer, K.M., and Pasternak, G.W. 1995. Differential blockade of morphine and morphine-6β-glucuronide analgesia by antisense oligodeoxynucleotides directed against MOR-1 and G-protein α subunits in rats. *Neurosci Lett* 198:99–102.

Rossi, G., Pan, Y.X., Cheng, J., and Pasternak, G.W. 1994. Blockade of morphine analgesia by an antisense oligodeoxynucleotide against the μ receptor. *Life Sci* 54:PL375–79.

Rossi, G., Perlmutter, M., Leventhal, L., Talatti, A., and Pasternak, G. 1998. Orphanin FQ/nociceptin analgesia in the rat. *Brain Res* 792:327–30.

Rossignol, S. 1996. Neural control of stereotypic limb movements. In *Handbook of Physiology, Section 12. Exercise: Regulation and Integration of Multiple Systems,* ed. L. Rowell and J. Sheperd, 173–216. Washington, D.C.: American Physiological Society.

Rothfield, J.M., Gross, D.S., and Watkins, L.R. 1985. Sexual responsiveness and its relationship to vaginal stimulation-produced analgesia in the rat. *Brain Res* 358:309–15.

Roychowdhury, S.M., and Fields, H.L. 1996. Endogenous opioids acting at a medullary μ-opioid receptor contribute to the behavioral antinociception produced by GABA antagonism in the midbrain periaqueductal gray. *Neuroscience* 74:863–72.

Roychowdhury, S.M., and Heinricher, M.M. 1997. Effects of iontophoretically applied serotonin on three classes of physiologically characterized putative pain modulating neurons in the rostral ventromedial medulla of lightly anesthetized rats. *Neurosci Lett* 226:136–38.

Rubinow, D., Schmidt, P., and Roca, C. 1998. Estrogen-serotonin interactions: implications for affective regulation. *Biol Psychiatry* 44:839–50.

Ryan, S.M., and Maier, S.F. 1988. The estrous cycle and estrogen modulate stress-induced analgesia. *Behav Neurosci* 102:371–80.

Rye, D.B., Lee, H.J., Saper, C.B., and Wainer, B.H. 1988. Medullary and spinal efferents of the pedunculopontine tegmental nucleus and adjacent meso-pontine tegmentum in the rat. *J Comp Neurol* 269:315–41.

Saksida, L., Galea, L., and Kavaliers, M. 1993. Predator-induced opioid and non-opioid mediated analgesia in young meadow voles: sex differences and de-velopmental changes. *Brain Res* 617:214–19.

Sakuma, Y., and Pfaff, D. 1980. LH-RH in the mesencephalic central gray can potentiate lordosis reflex of female rats. *Nature* 283:566–67.

Sakuma, Y., and Pfaff, D.W. 1979a. Facilitation of female reproductive behavior from mesencephalic central gray in the rat. *Am J Physiol* 237:R278–84.

———. 1979b. Mesencephalic mechanisms for integration of female reproduc-tive behavior in the rat. *Am J Physiol* 237:R285–90.

Sakurada, S., Zadina, J.E., Kastin, A.J., Katsuyama, S., Fujimura, T., Murayama, K., Yuki, M., Ueda, H., and Sakurada, T. 1999. Differential involvement of μ-opioid receptor subtypes in endomorphin-1 and-2-induced antinocicep-tion. *Eur J Pharmacol* 372:25–30.

Saleh, T., and Connell, B. 1998. Role of 17β-estradiol in the modulation of baro-reflex sensitivity in male rats. *Am J Physiol* 275:R770–78.

Samanin, R., and Valzelli, L. 1971. Increase of morphine-induced analgesia by stimulation of the nucleus raphe dorsalis. *Eur J Pharmacol* 16:298–302.

Samanin, R., Gulmulka, W., and Valzelli, L. 1970. Reduced effect of morphine in midbrain raphe lesioned rats. *Eur J Pharmacol* 10:339–43.

Sander, H.W., Kream, R.M., and Gintzler, A.R. 1989. Spinal dynorphin involve-ment in the analgesia of pregnancy: effects of intrathecal dynorphin antis-era. *Eur J Pharmacol* 159:205–9.

Sander, H.W., Portoghese, P.S., and Gintzler, A.R. 1988. Spinal κ opiate recep-tor involvement in the analgesia of pregnancy: effects of intrathecal nor-binaltorphamine, a κ-selective antagonist. *Brain Res* 474:343–47.

Sandkuhler, J., Maisch, B., and Zimmermann, M. 1987. Raphe magnus-induced descending inhibition of spinal nociceptive neurons is mediated through contralateral spinal pathways in the cat. *Neurosci Lett* 76:168–72.

Sanghera, M., Anselmo-Franci, J., and McCann, S. 1991. Effect of medial zona incerta lesions on the ovulatory surge of gonadotrophins and prolactin in the rat. *Neuroendocrinology* 54:433–38.

Saper, C.B., Swanson, L.W., and Cowan, W.M. 1976. The efferent connections of the ventromedial nucleus of the hypothalamus of the rat. *J Comp Neurol* 169(4):409–42.

Sar, M. 1984. Estradiol is concentrated in tyrosine hydroxylase-containing neu-rons of the hypothalamus. *Science* 223:938–40.

Sar, M., Stumpf, W.E., Miller, R.J., Chang, K.J., and Cuatrecasas, P. 1978. Im-munohistochemical localization of enkephalin in rat brain and spinal cord. *J Comp Neurol* 182:17–37.

Sarrel, P. 1990. Ovarian hormones and the circulation. *Maturitas* 590:287–98.

Satoh, M., and Takagi, H. 1971. Enhancement by morphine of the central descending inhibitory influence on spinal sensory transmission. *Eur J Pharmacol* 14:60–65.

Satoh, M., Oku, R., and Akaike, A. 1983. Analgesia produced by microinjection of L-glutamate into the rostral ventromedial bulbar nuclei of the rat and its inhibition by intrathecal α-adrenergic blocking agents. *Brain Res* 261:361–64.

Sato, M., and Yamanouchi, K. 1999. Efferent projections of hypothalamic neurons containing estrogen receptors α to midbrain periaqueductal gray: a retrograde tracing study using cholera toxin B subunit in rats. *Soc Neurosci* 25(abstr. 167.15):420.

Saunders, D.R., Paolino, R.M., Bosquet, W.F., and Miya, T.S. 1974. Age-related responsiveness of the rat to drugs affecting the central nervous system. *Proc J Exp Biol Med* 147:593–95.

Sawyer, C.H., and Kawakami, M. 1961. Interactions between the central nervous system and hormones influencing ovulation. In *Control of Ovulation*, ed. C.A. Villee, 79–97. New York: Pergamon.

Schmauss, C., and Yaksh, T.L. 1984. In vivo studies on spinal opiate receptor systems mediating antinociception. II. Pharmacological profiles suggesting a differential association of μ, δ and κ receptors with visceral chemical and cutaneous thermal stimuli in the rat. *J Pharmacol Exp Ther* 228:1–12.

Schmidt, L., and Schulkin, J., eds. 1999. *Extreme Fear, Shyness, and Social Phobia.* New York: Oxford University Press.

Schubert, K., Shaikh, M., and Siegel, A. 1996. NMDA receptors in the midbrain periaqueductal gray mediate hypothalamically evolved hissing behavior in the cat. *Brain Res* 762:80–90.

Schuller, A.G., King, M.A., Zhang, J., et al. 1999. Retention of heroin and morphine-6 β-glucuronide analgesia in a new line of mice lacking exon 1 of MOR-1. *Nat Neurosci* 2:151–56.

Schultz, W., Dayan, P., and Montague, P. 1997. A neural substrate of prediction and reward. *Science* 275:1593–99.

Schulz, S., Schreff, M., Koch, T., Zimprich, A., Gramsch, C., Elde, R., and Hollt, V. 1998. Immunolocalization of two μ-opioid receptor isoforms (MOR1 and MOR1B) in the rat central nervous system. *Neuroscience* 82:613–22.

Schwartz-Giblin, S., Korotzer, A., and Pfaff, D. 1989. Steroid hormone effects on picrotoxin-induced seizures in female and male rats. *Brain Res* 476:240–47.

Scott, C., Rawson, J., Pereira, A., and Clarke, I. 1999. Estrogen receptors in the brainstem of the female sheep: relationship to noradrenergic cells and cells projecting to the medial preoptic area. *J Neuroendocrinol* 11:503–12.

Segal, M., and Sandberg, D. 1977. Analgesia produced by electrical stimulation of catecholamine nuclei in the rat brain. *Brain Res* 123:369–72.

Selye, H. 1952. *The Story of the Adaptation Syndrome.* Montreal: Acta.

Servos, P., Barke, K., Hough, L., and Vanderwolf, C. 1994. Histamine does not play an essential role in electrocortical activation during waking behavior. *Brain Res* 636:98–102.

Shaikh, M., Brutus, M., Siegel, H., and Siegel, A. 1985. Topographically organized midbrain modulation of predatory and defensive aggression in the cat. *Brain Res* 336:308–12.

Shaikh, M., DeLanerolle, N., and Siegel, A. 1997. Serotonin 5HT-1A and 5HT-2/1C receptors in the midbrain periaqueductal gray differentially modulate defensive rage behavior elicited from the medial hypothalamus of the cat. *Brain Res* 765:198–207.

Shaikh, M., Lu, C.-L., and Siegel, A. 1991a. Affective defense behavior elicited from the feline midbrain periaqueductal gray is regulated by μ and δ opioid receptors. *Brain Res* 557:344–48.

———. 1991b. An enkephalinergic mechanism involved in amygdaloid suppression of affective defense behavior elicited from the midbrain periaqueductal gray in the cat. *Brain Res* 559:109–17.

Shaikh, M., Schubert, K., and Siegel, A. 1994. Basal amygdaloid facilitation of midbrain periaqueductal gray elicited defensive rage beahavior in the cat is mediated through NMDA receptors. *Brain Res* 635:187–95.

Shaikh, M., Steinberg, A., and Siegel, A. 1993. Evidence that substance P is utilized in medial amygdaloid facilitation of defensive rage behavior in the cat. *Brain Res* 625:283–94.

Shane, R., Lazar, D., Rossi, G., Pasternak, G., and Bodnar, R. 2000. Multiple opioid receptor subtype antagonists affect orphanin FQ/nociceptin analgesia in the rat amygdala. *Soc Neurosci Abstr* 6:436.

Shane, R., Rossi, G., Allen, R., Mathis, J., Pasternak, G., and Bodnar, R. 1999. Analgesic actions of orphanin FQ/nociceptin fragments in the ventrolateral periaqueductal gray and amygdala in rats. *Soc Neurosci Abstr* 25:1438.

Shane, R., Wilk, S., and Bodnar, R.J. 1999. Modulation of endomorphin-2-induced analgesia by dipeptidyl peptidase IV. *Brain Res* 815:278–86.

Shen, K., Crain, S., and Ledeen, R. 1991. Brief treatment of sensory ganglion neurons with GM1 ganglioside enhances the efficacy of opioid excitatory effects on the action potential. *Brain Res* 559:130–38.

Sherin, J., Shiromani, P., McCarley, R., and Saper, C. 1996. Activation of ventrolateral preoptic neurons during sleep. *Science* 271:216–19.

Sherman, J.E., Strub, H., and Lewis, J.W. 1984. Morphine analgesia: enhancement by shock-associated cues. *Behav Neurosci* 98:293–309.

Sherwin, B. 1998. Estrogen and cognitive functioning in women. *Estrogen Cogn* 217:17–22.

Shughrue, P., Komm, B., and Merchenthaler, I. 1996. The distribution of estrogen receptor β mRNA in the rat hypothalamus. *Steroids* 61:678–81.

Siegel, A., Schubert, K., and Shaikh, M. 1997. Neurotransmitters regulating defensive rage behavior in the cat. *Neurosci Biobehav Rev* 21:733–42.

Siegel, J.M., Nienhuis, R., Fahringer, H.M., Paul, R., Shiromani, P., Dement, W.C., Mignot, E., and Chiu, C. 1991. Neuronal activity in narcolepsy: identification of cataplexy-related cells in the medial medulla. *Science* 252:1315–18.

Siegfried, B., and Nunes de Souza, R.L. 1989. NMDA receptor blockade in the periaqueductal grey prevents stress-induced analgesia in attacked mice. *Eur J Pharmacol* 168:239–42.

Siegfried, B., Frischnecht, H.R., and Waser, P.G. 1984. Defeat, learned submissiveness and analgesia: effect of genotype. *Behav Neural Biol* 42:91–97.

Siegfried, B., Freischnecht, H.R., Riggio, G., and Waser, P.G. 1987. Long-term analgesic reaction in attacked mice. *Behav Neurosci* 101:797.

Siggins, G.R., Henriksen, S.J., Chavkin, C., and Gruol, D. 1986. Opioid peptides and epileptogenesis in the limbic system: cellular mechanisms. *Adv Neurol* 44(1):501–12.

Simon, E.J., Hiller, J.M., and Edelman, I. 1973. Stereospecific binding of the potent narcotic analgesic ($^3$H)etorphine to rat brain homogenate. *Proc Natl Acad Sci USA* 70:1947–49.

Simone, D.A., and Bodnar, R.J. 1982. Modulation of antinociceptive responses following tail pinch stress. *Life Sci* 30:719–29.

Simonian, S., and Herbison, A. 1997. Differential expression of estrogen receptor and neuropeptide Y by brainstem A1 and A2 noradrenaline neurons. *Neuroscience* 76:517–29.

Simonian, S., Delaleu, B., Caraty, A., and Herbison, A. 1998. Estrogen receptor expression in brainstem noradrenergic neurons of the sheep. *Neuroendocrinology* 67:392–402.

Siuciak, J.A., and Advokat, C. 1989. The synergistic effects of concurrent spinal and supraspinal opiate agonisms is reduced by both nociceptive and morphine pretreatment. *Pharmacol Biochem Behav* 34:265–73.

Skinner, K., Fields, H.L., Basbaum, A.I., and Mason, P. 1997. GABA-immunoreactive boutons contact identified OFF and ON cells in the nucleus raphe magnus. *J Comp Neurol* 378:196–204.

Slowe, S.J., Simonin, F., Kieffer, B., and Kitchen, I. 1999. Quantitative autoradiography of μ-, δ- and κ1 opioid receptors in κ-opioid receptor knockout mice. *Brain Res* 818:335–45.

Smith, D.J., Perotti, J.M., Crisp, T., Cabral, M.E.Y., Long, J.T., and Scalziti, J.N. 1988. The mu receptor is responsible for descending pain inhibition originating in the periaqueductal gray region of the rat brain. *Eur J Pharmacol* 156:47–54.

Smith, D.J., Robertson, B., and Monroe, P.J. 1992. Antinociception from the administration of β-endorphin into the periaqueductal gray is enhanced while that of morphine is inhibited by barbiturate anesthesia. *Neurosci Lett* 146:143–46.

Smith, D.J., Robertson, B., Monroe, P.J., Leedham, J.A., and Cabral, J.D.Y. 1992. Opioid receptors mediating antinociception from β-endorphin and morphine in the periaqueductal gray. *Neuropharmacology* 31:1137–50.

Smith, D.W., and Day, T.A. 1994. C-Fos expression in hypothalamic neurosecretory and brainstem catecholamine cells following noxious somatic stimuli. *Neuroscience* 58:765–75.

Smith, S.S. 1997. Estrous hormones enhance coupled, rhythmic olivary discharge in correlation with facilitated limb stepping. *Neuroscience* 82:83–95.

Sora, I., Funada, M., and Uhl, G. 1997. The μ-opioid receptor is necessary for [D-Pen2, D-Pen5] enkephalin-induced analgesia. *Eur J Pharmacol* 324:R1–2.

Sperber, E.S., Kramer, E., and Bodnar, R.J. 1986. Effects of muscarinic receptor antagonism upon two forms of stress-induced analgesia. *Pharmacol Biochem Behav* 25:171–79.

Spiaggia, A., Bodnar, R.J., Kelly, D.D., and Glusman, M. 1979. Opiate and nonopiate mechanisms of stress-induced analgesia: cross-tolerance between stressors. *Pharmacol Biochem Behav* 10:761–65.

Spiegel, K., Kourides, I., and Pasternak, G.W. 1982. Prolactin and growth hormone release by morphine in the brain: different receptor mechanisms. *Science* 217:745–47.

Spinella, M., and Bodnar, R.J. 1994. Nitric oxide synthase inhibition selectively potentiates swim stress antinociception in rats. *Pharmacol Biochem Behav* 47: 727–33.

Spinella, M., Cooper, M.L., and Bodnar, R.J. 1996. Excitatory amino acid antagonists in the rostral ventromedial medulla inhibit mesencephalic morphine analgesia in rats. *Pain* 64:545–52.

Spinella, M., Schaefer, L.A., and Bodnar, R.J. 1997. Ventral medullary mediation of mesencephalic morphine analgesia by muscarinic and nicotinic cholinergic receptor antagonists in rats. *Analgesia* 3:119–30.

Spinella, M., Znamensky, V., Moroz, M., Ragnauth, A., and Bodnar, R.J. 1999. Actions of NMDA and cholinergic receptor antagonists in the rostral ventromedial medulla upon β-endorphin analgesia elicited from the ventrolateral periaqueductal gray. *Brain Res* 829:151–59.

Sprague, J.M., Chambers, W.W., and Stellar, E. 1961. Attentive, affective and adaptive behavior in the cat. *Science* 133:165–73.

Spratto, G.R., and Dorio, R.E. 1978. Effect of age on acute morphine response in the rat. *Res Commun Chem Pathol Pharmacol* 19:23–26.

Stallings, M., and Hewitt, J. 1996. Genetic and environmental structure of the tridimensional personality questionnaire: three or four temperament dimensions. *J Pers Social Psychol* 70:127–40.

Standifer, K., Chien, C., Wahlestedt, C., Brown, G., and Pasternak, G. 1994. Selective loss of δ opioid analgesia and binding by antisense oligodeoxynucleotides to a δ opioid receptor. *Neuron* 12:805–10.

Standifer, K.M., Jenab, S., Su, W., Chien, C.-C., Pan, Y.-X., Inturrisi, C.E., and Pasternak, G.W. 1995. Antisense oligodeoxynucleotides to the cloned receptor, DOR-1: uptake, stability and regulation of gene expression. *J Neurochem* 65:1981–87.

Steinfels, G.F., Heym, J., Strecker, R.E., and Jacobs, B.L. 1983. Behavioral correlates of dopaminergic unit activity in freely moving cats. *Brain Res* 258:217–28.

Steinman, J.L., Faris, P.L., Mann, P.E., Olney, J.W., Komisaruk, B.R., Willis, W.D., and Bodnar, R.J. 1990. Antagonism of morphine analgesia by nonopioid cold-water swim analgesia: direct evidence for collateral inhibition. *Neurosci Biobehav Rev* 14:1–7.

Steinman, J.L., Komisaruk, B.R., Yaksh, T.L., and Tyce, G.M. 1983. Spinal cord monoamines mediate the antinociceptive effects of vaginal stimulation in rats. *Pain* 16:155–66.

Stellar, E. 1954. The physiology of motivation. *Psychol Rev* 61:5–31.

Steriade, M. 1996. Arousal: revisiting the reticular activating system. *Science* 272:225–26.

Steriade, M., Contreras, D., and Amzica, F. 1997. The thalamocortical dialogue during wake, sleep, and paroxysmal oscillations. In *Thalamus*, vol. 2, ed. M. Steriade, E. Jones, and D. McCormick, 213–94. Amsterdam: Elsevier.

Sternberg, W.F., Mogil, J.S., Kest, B., Page, G.G., Leong, Y., Yam, V., and Liebeskind, J.C. 1995. Neonatal testosterone exposure affects neurochemistry of non-opioid swim stress-induced analgesia in adult mice. *Pain* 63:321–26.

Sternini, C., Spann, M., Anton, B., Keith, D.E., Jr., Bunnett, N.W., von Zastrow, M., Evans, C., and Brecha, N.C. 1996. Agonist-selective endocytosis of μ opioid receptor by neurons in vivo. *Proc Natl Acad Sci USA* 93(17):9241–46.

Stewart, D., MacFabe, D., and Vanderwolf, C. 1984. Cholinergic activation of the electrocorticogram: role of the substantia innominata and effects of atropine and quinuclidinyl benzilate. *Brain Res* 322:219–32.

Stoleru, S., Redoute, I., Gregoire, M.-E., Lavenne, F., Le Bars, D., Cinotti, L., Spira, A., and Pujol, J. 1998. Cerebral correlates of hypoactive sexual desire disorder. *Soc Neurosci Abstr* 24(abstr. 368.2):934.

Stone, L.S., Fairbanks, C.A., Laughlin, T.M., Nguyen, H.O., Bushy, T.M., Wessendorf, M.W., and Wilcox, G.L. 1997. Spinal analgesic actions of the new endogenous opioid peptides endomorphin-1 and-2. *Neuroreport* 8:3131–35.

Stricker, E., and Zigmond, M. 1984. In *Cetacholamines*, Part B, ed. E. Usdin, A. Carlsson, and A. Dahlstrom, 259–69. New York: Liss.

Suh, H.H., Fujimoto, J.M., and Tseng, L.F. 1989. Differential mechanisms mediating β-endorphin-and morphine-induced analgesia in mice. *Eur J Pharmacol* 168:61–70.

Suh, H.H., and Tseng, L.F. 1990. Differential effects of sulfated cholecystokinin octapeptide and proglumide injected intrathecally on antinociception induced by β-endorphin and morphine administered intracerebroventricularly in mice. *Eur J Pharmacol* 179:329–38.

Sullivan, A.F., McQuay, H.J., Bailey, D., and Dickenson, A.H. 1989. The spinal antinociceptive actions of morphine metabolites, morphine-6-glucuronide and normorphine in the rat. *Brain Res* 482:219–24.

Sumner, B., and Fink, G. 1997. The density of 5-hydroxytryptamine$_{2A}$ receptors in forebrain is increased at pro-oestrus in intact female rats. *Neurosci Lett* 234:7–10.

————. 1998. Testosterone as well as estrogen increases serotonin$_{2A}$ receptor mRNA and binding site densities in the male rat brain. *Mol Brain Res* 59: 205–14.

Svoboda, K., and Lupica, C. 1998. Opioid inhibition of hippocampal interneurons via modulation of potassium and hyperpolarization-activated cation ($I_h$) currents. *J Neurosci* 18:7084–98.

Svrakic, D., Przybeck, T., and Cloninger, C. 1992. Mood states and personality traits. *J Affect Disord* 24:217–26.

Svrakic, D., Whitehead, C., Przybeck, T., and Cloninger, C. 1993. Differential diagnosis of personality disorders by the seven-factor model of temperament and character. *Arch Gen Psychiatry* 50:991–99.

Swanson, L.W., and Kuypers, H.G.J.M. 1980. The paraventricular nucleus of the hypothalamus: cytoarchitectonic subdivisions and the organization of projections to the pituitary, dorsal vagal complex and spinal cord as demonstrated by retrograde fluorescence double-labeling methods. *J Comp Neurol* 194:555–70.

Swanson, L.W., and Sawchenko, P.E. 1983. Hypothalamic integration: organization of the paraventricular and supraoptic nuclei. *Annu Rev Neurosci* 6:269–324.

Swanson, L.W., Sawchenko, P.E., Berod, A., Hartman, B.K., Helle, K.B., and Vanorden, D.E. 1981. An immunohistochemical study of the organization of catecholaminergic cells and terminal fields in the paraventricular and supraoptic nuclei of the hypothalamus. *J Comp Neurol* 196:271–85.

Takagi, H., Satoh, M., Akaike, A., Shibata, T., and Kuraishi, Y. 1977. The nucleus reticularis gigantocellularis of the medulla oblongata is a highly sensitive site in the production of morphine analgesia in the rat. *Eur J Pharmacol* 45: 91–92.

Tenen, S.S. 1968. Antagonism of the analgesic effect of morphine and other drugs by p-chlorophenylalanine, a serotonin depletor. *Psychopharmacology* 12:278–85.

Terenius, L. 1973. Stereospecific interaction between narcotic analgesia and a synaptic plasma membrane fraction of rat cerebral cortex. *Acta Pharmacologica Toxicilogica* 32:317–20.

Terman, G.W., Morgan, M.J., and Liebeskind, J.C. 1986. Opioid and nonopioid stress analgesia from cold water swims: importance of stress severity. *Brain Res* 372:167–71.

Terman, G.W., Penner, E.R., and Liebeskind, J.C. 1985. Stimulation-produced and stress-induced analgesia: cross-tolerance between opioid forms. *Brain Res* 360:374–78.

Terman, G.W., Shavit, Y., Lewis, J.W., Cannon, J.T., and Liebeskind, J.C. 1984. Intrinsic mechanisms of pain inhibition: activation by stress. *Science* 226: 1270–77.

Tershner, S.A., and Helmstetter, F.J. 1995. Spinal antinociception following stimulation of the amygdala depends on opioid receptors in the ventral peri-aqueductal gray. *Soc Neurosci Abstr* 21:1169.

Tershner, S., Mitchell, J., and Fields, H. 2000. Brainstem pain modulating circuitry is sexually dimorphic with respect to μ and κ opioid receptor function. *Pain* 85:153–59.

Teskey, G.C., Kavaliers, M., and Hirst, M. 1984. Social conflict activates opioid analgesic and ingestive behaviors in male mice. *Life Sci* 35:303–15.

Teyke, T., Weiss, K.R., and Kupfermann, I. 1990. An identified neuron (CPR) evokes neuronal responses reflecting food arousal in aplysia. *Science* 24:85–87.

Teyke, T., Xin, Y., Weiss, K., and Kupfermann, I. 1997. Ganglionic distribution of inputs and outputs of C-PR, a neuron involved in the generation of a food-induced arousal state in aplysia. *Invertebr Neurosci* 2:235–44.

Thayer, R. 1970. Activation states as assessed by verbal report and four psychophysiological variables. *Psychophysiology* 7:86–94.

Thompson, M.L., Miczek, K., Noda, K., Schuster, L., and Kumar, M.S.A. 1988. Analgesia in defeated mice: evidence for mediation via central rather than pituitary or adrenal endogenous opioid peptides. *Pharmacol Biochem Behav* 29:451–56.

Tian, J.H., Xu, W., Fang, Y., Mogil, J.S., Grisel, J.E., Grandy, D.K., and Han, J.S. 1997. Bidirectional modulatory effect of orphanin FQ on morphine-induced analgesia: antagonism in brain and potentiation in spinal cord of the rat. *Br J Pharmacol* 120:676–80.

Tinbergen, N. 1951. *The Study of Instinct*. Oxford: Clarendon.

Torii, M., Kubo, K., and Sasaki, T. 1995. Naloxone and initial estrogen action to induce lordosis in ovariectomized rats: the effect of a cut between the septum and preoptic area. *Neurosci Lett* 195:167–70.

————. 1996. Influence of opioid peptides on the priming action of estrogen on lordosis in ovariectomized rats. *Neurosci Lett* 212:68–70.

————. 1997. Differential effects of β-endorphin and met-and leu-enkephalin on steroid hormone-induced lordosis in ovariectomized female rats. *Pharmacol Biochem Behav* 58:837–42.

Trevino, A., Wolf, A., Jackson, A., Price, T., and Uphouse, L. 1999. Reduced efficacy of 8-OH-DPAT's inhibition of lordosis behavior by prior estrogen treatment. *Horm Behav* 35:215–22.

Trujillo, K.A., and Akil, H. 1991. Inhibition of morphine tolerance and dependence by the NMDA receptor antagonist, MK-801. *Science* 251:85–87.

Tseng, L.F. 1995. Mechanisms of β-endorphin-induced antinociception. In *The Pharmacology of Opioid Peptides*, ed. L. F. Tseng, 249. Chur, Switzerland: Hardwood.

Tseng, L.F., and Collins, K.A. 1991. Different mechanisms mediating tail-flick inhibition induced by β-endorphin, DAMGO and morphine from ROb and GiA in anesthetized rats. *J Pharmacol Exp Ther* 257:530–38.

————. 1992. Cholecystokinin administered intrathecally selectively antagonizes intracerebroventricular β-endorphin-induced tail-flick inhibition in the mouse. *J Pharmacol Exp Ther* 260:1086–92.

Tseng, L.F., and Fujimoto, J.M. 1985. Differential actions of intrathecal naloxone on blocking the tail-flick inhibition induced by intraventricular β-endorphin and morphine in rats. *J Pharmacol Exp Ther* 232:74–79.

Tseng, L.F., and Suh, H.H. 1989. Intrathecal [Met5]-enkephalin antibody blocks analgesia induced by intracerebroventricular β-endorphin but not morphine in mice. *Eur J Pharmacol* 173:171–76.

Tseng, L.F., and Tang, R. 1990. Different mechanisms mediate β-endorphin-and morphine-induced inhibition of the tail-flick response in rats. *J Pharmacol Exp Ther* 252:546–51.

Tseng, L.F., and Wang, Q. 1992. Forebrain sites differentially sensitive to β-endorphin and morphine for analgesia and release of met-enkephalin in the pentobarbital-anesthetized rat. *J Pharmacol Exp Ther* 261:1028–36.

Tseng, L.F., Collins, K.A., and Kampine, J.P. 1994. Antisense oligodeoxynucleotides to a δ-opioid receptor selectively blocks the spinal antinociception induced by δ-, but not μ- or κ-opioid receptor agonists in the mouse. *Eur J Pharmacol* 258:R1–3.

Tseng, L.F., Higgins, M.J., Hong, J.S., Hudson, P.M., and Fujimoto, J.M. 1985. Release of immunoreactive met-enkephalin from the spinal cord by intraventricular β-endorphin but not morphine in anesthetized rats. *Brain Res* 343:60–69.

Tseng, L.F., Ostwald, T.J., Loh, H.H., and Li, C.H. 1979. Behavioral activities of opioid peptides and morphine sulfate in golden hamsters and rats. *Psychopharmacology* 64:215–18.

Tseng, L.F., Suganuma, C., Narita, M., Oji, G.S., Tseng, E.L., Bhaita, A., Mizoguchi, H., and Nagase, H. 1998. Differential mechanism mediating antinociception induced by supraspinally administered newly discovered endogenous opioid peptides endomorphin-1 and endomorphin-2 in the mouse. *Soc Neurosci Abstr* 24:1356.

Tsou, K., and Jang, C.S. 1964. Studies on the site of analgesic action of morphine by intracerebral microinjection. *Scientia Sinica* 7:1099–1109.

Turcotte, J.C., and Blaustein, J.D. 1999. Projections of the estrogen receptor-immunoreactive ventrolateral hypothalamus to other estrogen receptor-immunoreactive sites in female guinea pig brain. *Neuroendocrinology* 69:63–76.

Uhl, G.R., Childers, S.R., and Pasternak, G.W. 1994. An opiate receptor gene family reunion. *Trends Neurosci* 17:89–93.

Unruh, A. 1996. Gender variations in clinical pain experience. *Pain* 65:123–67.

Unruh, A., Ritchie, J., and Merskey, H. 1999. Does gender affect appraisal of pain and pain coping strategies? *Clin J Pain* 15:31–40.

Uphouse, L., Andrade, M., Caldarola-Pastuszka, M., and Jackson, A. 1996a. 5-HT1A receptor antagonists and lordosis behavior. *Neuropharmacology* 35:489–95.

Uphouse, L., Colon, L., Cox, A., Caldarola-Pastuszka, M., and Wolf, A. 1996b. Effects of mianserin and ketanserin on lordosis behavior after systemic treatment or infusion into the ventromedial nucleus of the hypothalamus. *Brain Res* 718:46–52.

Urban, M.O., and Smith, D.J. 1993. Role of neurotensin in the nucleus raphe magnus in opioid-induced antinociception from the periaqueductal gray. *J Pharmacol Exp Ther* 265:580–86.

Urban, M.O., and Smith, D.J. 1994a. Nuclei within the rostral ventromedial medulla mediating morphine antinociception from the periaqueductal gray. *Brain Res* 652:9–16.

———. 1994b. Localization of the antinociceptive and antianalgesic effects of neurotensin within the rostral ventromedial medulla. *Neurosci Lett* 174:21–25.

Usher, M., Cohen, J., Servan-Schreiber, D., Rajkowski, J., and Aston-Jones, G. 1999. The role of locus coeruleus in the regulation of cognitive performance. *Science* 283:549–54.

Vaccarino, A.L., Marek, P., Sternberg, W., and Liebeskind, J.C. 1992. NMDA receptor antagonist MK-801 blocks nonopioid stress-induced analgesia in the formalin test. *Pain* 50:119–23.

Valenstein, E.S., Cox, V.C., and Kakolewski, J.W. 1970. Reexamination of the role of the hypothalamus in motivation. *Psychol Rev* 77:16–31.

Valentino, R., Foote, S., and Aston-Jones, G. 1983. Corticotropin-releasing factor activates noradrenergic neurons of the locus coeruleus. *Brain Res* 270: 363–67.

Valentino, R., Page, M., and Curtis, A. 1991. Activation of noradrenergic locus coeruleus neurons by hemodynamic stress is due to local release of corticotropin-releasing factor. *Brain Res* 555:25–34.

VanBockstaele, E., Colago, E., and Valentino, R. 1998. Amygdaloid corticotropin-releasing factor targets locus coeruleus dendrites: substrate for the coordination of emotional and cognitive limbs of the stress response. *J Neuroendocrinol* 10:743–57.

VanBockstaele, E.J., Aston-Jones, G., and Pierbone, V.A. 1991. Subregions of the periaqueductal gray topographically innervate the rostral ventral medulla in the rat. *J Comp Neurol* 309:305–27.

VanBockstaele, E.J., Pieribone, V.A., and Aston-Jones, G. 1989. Diverse afferents converge on the nucleus paragigantocellularis in the rat ventrolateral medulla: retrograde and anterograde tracing studies. *J Comp Neurol* 290:561–84.

Vanderah, T.W., Wild, K.D., Takemori, A.E., Sultana, M., Portoghese, P.S., Bowen, W.D., Mosberg, H.I., and Porreca, F. 1992. Mediation of swim-stress antinociception by the δ-2 receptor in the mouse. *J Pharmacol Exp Ther* 262:190–97.

Vanderhorst, V.G.J.M., Schasfoort, F.C., Meijer, E., Van Leeuwen, F.W., and Holstege, G. 1998. Estrogen receptor-δ-immunoreactive neurons in the peri-aqueductal gray of the adult ovariectomized female cat. *Neurosci Lett* 240: 13–16.

Vanderschuren, L., Wardeh, G., DeVries, T., Mulder, A., and Schoffelmeer, A. 1999. Opposing role of dopamine D1 and D2 receptors in modulation of rat nucleus accumbens noradrenaline release. *J Neurosci* 19:4123–31.

Vanderwolf, C. 1884. Aminergic control of the electrocorticogram: a progress report. In *Neurobiology of the Trace Amines,* ed. A. Boulton, G. Baker, W. Dewhurst, and M. Sandler, 163–83. Clifton, N.J.: Humana.

———. 1988. Cerebral activity and behavior: control by central cholinergic and serotonergic systems. In *International Review of Neurobiology,* vol. 30, 225–340. San Diego: Academic.

Vanderwolf, C., and Baker, G. 1996. The role of brain noradrenaline in cortical activation and behavior: a study of lesions of the locus coeruleus, medial thalamus and hippocampus-neocortex and of muscarinic blockade in the rat. *Behav Brain Res* 78:225–34.

Vanderwolf, C., and Stewart, D. 1986. Joint cholinergic-serotonergic control of neocortical and hippocampal electrical activity in relation to behavior: effects of scopolamine, ditran, trifluoperazine and amphetamine. *Physiol Behav* 38:57–65.

Vanderwolf, C., Leung, L., Baker, G., and Stewart, D. 1989. The role of serotonin in the control of cerebral activity: studies with intracerebral 5,7-dihydroxytryptamine. *Brain Res* 504:181–91.

Vanegas, H., and Barbaro, N. 1984. Tail-flick related activity in medullospinal neurons. *Brain Res* 321:135–41.

vanPraag, H., and Frenk, H. 1990. The role of glutamate in opiate descending inhibition of nociceptive spinal reflexes. *Brain Res* 524:101–5.

VanVoigtlander, P.F., Lahti, R.A., and Ludens, J.H. 1983. U50488H: a selective and structurally novel non-μ (κ) opioid agonist. *J Pharmacol Exp Ther* 224:7–11.

Vaupel, B. 1983. Naloxone fails to antagonize the delta effects of PCP and SKF 10,047 in the dog. *Eur J Pharmacol* 92:269–74.

Vogt, M. 1974. The effect of lowering the 5-hydroxytryptamine content of the rat spinal cord on analgesia produced by morphine. *J Physiol* 236:483–98.

Wade, G., and Zucker, I. 1970. Modulation of food intake and locomotor activity in female rats by diencephalic hormone implants. *J Comp Physiol Psychol* 72:328–36.

Waeber, C., Dixon, K., Hoyer, D., and Palacios, J.M. 1988. Localization by autoradiography of 5HT-3 receptors in the mouse CNS. *Eur J Pharmacol* 151: 351–52.

Wahlestedt, C. 1994. Antisense oligodeoxynucleotide strategies in neuropharmacology. *Trends Pharmacol Sci* 15:42–46.

Wall, P. 1967. The laminar organization of dorsal horn and effects of descending impulses. *J Physiol (Lond)* 188:403–23.

Wang, G. 1923. The relation between "spontaneous" activity and oestrous cycle in the white rat. *Comp Psychol Monogr* 2:1–27.

Wang, L., and Gintzler, A. 1994. Bimodal opioid regulation of cyclic AMP formation: implications for positive and negative coupling of opiate receptors to adenylyl cyclase. *J Neurochem* 63:1726–30.

————. 1995. Morphine tolerance and physical dependence: reversal of opioid inhibition to enhancement of cyclic AMP formation. *J Neurochem* 64:1102–6.

Wardlaw, S.L., and France, A.G. 1983. Brain β-endorphin during pregnancy, parturition and the post-partum period. *Endocrinology* 113:1664–68.

Warner, L.H. 1927. A study of sex behavior in the white rat by means of the obstruction method. *Comp Psychol Monogr* 4:1–66.

Wasman, M., and Flynn, J. 1962. Directed attack elicited from the hypothalamus. *Arch Neurol* 6:220.

Watkins, L.R., and Mayer, D.J. 1982. Involvement of spinal opioid systems in footshock-induced analgesia: antagonism by naloxone is possible only before induction of analgesia. *Brain Res* 242:309–16.

Watkins, L.R., Cobelli, D.A., Faris, P., Aceto, M.D., and Mayer, D.J. 1982a. Opiate vs non-opiate footshock induced analgesia (FSIA): the body region shocked is a critical factor. *Brain Res* 242:299–308.

Watkins, L.R., Cobelli, D.A., Newsome, H.H., and Mayer, D.J. 1982b. Footshock induced analgesia is dependent neither on pituitary nor sympathetic activation. *Brain Res* 245:81–96.

Watkins, L.R., Drugan, R.C., Hyson, R.L., Moye, T.B., Ryan, S.M., Mayer, D.J., and Maier, S.F. 1984a. Opiate and nonopiate analgesia induced by inescapable tail shock: effects of dorsolateral funiculus lesions and decerebration. *Brain Res* 291:325–36.

Watkins, L.R., Faris, P.L., Komisaruk, B.R., and Mayer, D.J. 1984b. Dorsolateral funiculus and intraspinal pathways mediate vaginal stimulation-induced suppression of nociceptive responding in rats. *Brain Res* 294:59–65.

Watkins, L.R., Katayama, Y., Kinscheck, I.B., Mayer, D.J., and Hayes, R.L. 1984c. Muscarinic cholinergic mediation of opiate and nonopiate environmentally induced analgesias. *Brain Res* 300:231–42.

Watkins, L.R., Kinscheck, I.B., and Mayer, D.J. 1983. The neural basis of footshock analgesia: the effect of periaqueductal gray lesions and decerebration. *Brain Res* 276:317–24.

————. 1985. Potentiation of morphine analgesia by the cholecystokinin antagonist proglumide. *Brain Res* 327:169–80.

Watkins, L.R., Kinscheck, I.B., Kaufman, E.F.S., Miller, J., Frenk, H., and Mayer, D.J. 1985. Cholecystokinin antagonists selectively potentiate analgesia induced by endogenous opiates. *Brain Res* 327:181–90.

Watkins, L.R., Suberg, S.N., Thurston, C.L., and Culhane, E.S. 1986. Role of spinal cord neuropeptides in pain sensitivity and analgesia: thyrotropin releasing hormone and vasopressin. *Brain Res* 362:308–17.

Watkins, L.R., Wiertelak, E.P., and Maier, S.F. 1993. The amygdala is necessary for the expression of conditioned but not unconditioned analgesia. *Behav Neurosci* 107:402–5.

Watkins, L.R., Wiertelak, E.P., Grisel, J.E., Silbert, L.H., and Maier, S.F. 1992. Parallel activation of multiple spinal opiate systems appears to mediate 'non-opiate' stress-induced analgesias. *Brain Res* 594:99–108.

Watkins, L.R., Young, E.G., Kinscheck, I.B., and Mayer, D.J. 1983. The neural basis of footshock analgesia: the role of specific ventral medullary nuclei. *Brain Res* 276:305–15.

Watson, N., Hargreaves, E., Penava, D., Eckel, L., and Vanderwolf, C. 1992. Serotonin-dependent cerebral activation: effects of methiothepin and other serotonergic antagonists. *Brain Res* 597:16–23.

Watson, S.J., Akil, H., Richard, C.W., and Barchas, J.D. 1978. Evidence for two separate opiate peptide neuronal systems and the coexistence of β-lipotropin, β-endorphin and ACTH immunoreactivities in the same hypothalamic neurons. *Nature* 275:226–28.

Webster, G.W., Shuster, L., and Eleftheriou, B.E. 1976. Morphine analgesia in mice of different ages. *Exp Aging Res* 2:221–23.

Weissman, M., Livingston, B., and Leaf, P. 1991. Affective disorders. In *Psychiatric Disorders in America*, ed. L. Robins and D. Regier. New York: Free Press.

Westbrook, R.F., and Greeley, J.D. 1990. Some effects of the opioid antagonist naloxone upon the rat's reactions to a heat stressor. *Q J Exp Psychol* 42:1–40.

Whalen, R. 1974. Estrogen-progesterone induction of mating in female rats. *Horm Behav* 5:157–62.

Whipple, B., and Komisaruk, B.R. 1985. Elevation of pain threshold by vaginal stimulation in women. *Pain* 21:357–67.

Whipple, B., Josimovich, J.B., and Komisaruk, B.R. 1990. Sensory thresholds during the antepartum, intrapartum and postpartum periods. *Int J Nurs Stud* 27:213–21.

White, R., Darkow, D., and Lang, J. 1995. Estrogen relaxes coronary arteries by opening BKCa channels through a cGMP-dependent mechanism. *Circ Res* 77:936–42.

Wiklund, L., Behzadi, G., Kalen, P., Headley, P.M., Nicolopoulos, L.S., Parsons, C.G., and West, D.C. 1988. Autoradiographic and electrophysiological evidence for excitatory amino acid transmission in the periaqueductal gray projection to nucleus raphe magnus in the rat. *Neurosci Lett* 93:158–63.

Willer, J.C., and Ernst, M. 1986. Somatovegetative changes in stress-induced analgesia in man: an electrophysiological and pharmacological study. *Ann NY Acad Sci* 467:14.

Williams, F.G., and Beitz, A.J. 1990. Ultrastructural morphometric analysis of enkephalin-immunoreactive terminals in the ventrocaudal periaqueductal gray: analysis of their relationship to periaqueductal gray-raphe magnus projection neurons. *Neuroscience* 38(2):381–94.

Willis, W. 1985. *The Pain System: The Neural Basis of Nociceptive Transmission in the Mammalian Nervous System*. Basel, Switzerland: Karger.

———. 1986. Visceral inputs to sensory pathways in the spinal cord. *Prog Brain Res* 67:207–25.

———. 1988. Anatomy and physiology of descending control of nociceptive responses of dorsal horn neurons: comprehensive review. *Prog Brain Res* 77:1–29.

Wise, R. 1982. Neuroleptics and operant behavior: the anhedonia hypothesis. *Behav Brain Sci* 5:39–87.

Wise, R.A., and Bozarth, M. 1987. A psychomotor stimulant theory of addiction. *Psychol Rev* 94:469–92.

Wise, R., and Rompre, P. 1989. Brain dopamine and reward. *Annu Rev Psychol* 40:191–225.

Wittchen, H.-U., Essau, C., Von Zerssen, D., Krieg, J.-C., and Zaudig, M. 1992. Lifetime and six-month prevalence of mental disorders in the Munich follow-up study. *Eur Arch Psychiatry Clin Neurosci* 241:247–58.

Wong, C.L. 1987. The effect of gonadectomy on swim stress induced antinociception in mice. *Eur J Pharmacol* 142:159–61.

———. 1988. Effect of oestrodiol replacement on swim-induced antinociception in ovariectomized mice. *Clin Exp Pharmacol Physiol* 15:799–802.

Wright, D., and Jennes, L. 1993. Origin of noradrenergic projections to GnRH perikarya-containing areas in the medial septum-diagonal band and preoptic area. *Brain Res* 621:272–78.

Xiao, L., and Becker, J. 1998. Effects of estrogen agonists on amphetamine-stimulated striatal dopamine release. *Synapse* 29:379–91.

Yaksh, T.L. 1984a. Multiple opioid receptor systems in the brain and spinal cord. I. *Eur J Anesthesiol* 1:171–201.

———. 1984b. Multiple opioid receptor systems in the brain and spinal cord. II. *Eur J Anesthesiol* 1:201–43.

Yaksh, T.L., and Rudy, T.A. 1978. Narcotic analgetics: CNS sites on mechanisms of action as revealed by intracerebral injection techniques. *Pain* 4:299–359.

Yaksh, T.L., Yeung, J.A., and Rudy, T.A. 1976. Systematic examination in the rat of brain sites sensitive to the direct application of morphine: observations of differential effects within the periaqueductal gray. *Brain Res* 114:83–103.

Yamano, M., Inagaki, S., Kito, S., Matsuzaki, T., Shinohara, Y., and Tohyama, M. 1986. Enkephalinergic projection from the ventromedial hypothalamic nucleus to the midbrain central gray matter in the rat: an immunocytochemical analysis. *Brain Res* 398:337–46.

Yeomans, D.C., and Proudfit, H.K. 1992. Antinociception induced by microinjection of substance P into the A7 catecholamine cell group in the rat. *Neuroscience* 49:681–91.

Yeomans, D.C., Clark, F.M., Paice, J.A., and Proudfit, H.K. 1992. Antinociception induced by electrical stimulation of spinally-projecting noradrenergic neurons in the A7 catecholamine cell group of the rat. *Pain* 48:449–61.

Yeung, J.C., and Rudy, T.A. 1980. Multiplicative interaction between narcotic agonisms expressed at spinal and supraspinal sites of antinociceptive action as revealed by concurrent intrathecal and intracerebroventricular injections of morphine. *J Pharmacol Exp Ther* 215:633–42.

Yeung, J.C., Yaksh, T.L., and Rudy, T. 1977. Concurrent mapping of brain stem sites for sensitivity to the direct application of morphine and focal electrical stimulation in the production of antinociception in the rat. *Pain* 4:23–40.

Yin, J., Kaplitt, M., and Pfaff, D. 1995. In vivo promoter analysis in the adult central nervous system using viral vectors. In *Viral Vectors*, ed. M. Kaplitt and A. Loewy, 157–71. San Diego: Academic.

Yoburn, B.C., Truesdell, L.S., Kest, B., Inturrisi, C.R., and Bodnar, R.J. 1987. Chronic opioid antagonist treatment facilitates non-opioid stress-induced analgesia. *Pharmacol Biochem Behav* 27:525–27.

Yonkers, K., Bradshaw, K., and Halbreich, U. 2000. Oestrogens, progestins and mood. *Mood Disord Women:* 207–32.

Young, P.T. 1961. *Motivation and Emotion.* New York: John Wiley.

Zadina, J.E., Hackler, L., Ge, L.-J., and Kastin, A.J. 1997. A potent and selective endogenous agonist for the μ-opiate receptor. *Nature* 386:499–502.

Zardetto-Smith, A.M., and Gray, T.S. 1990. Organization of peptidergic and catecholaminergic efferents from the nucleus of the solitary tract to the rat amygdala. *Brain Res Bull* 25:875–87.

Zhang, S., Tong, Y., Tian, M., Dehaven, R.N., Cortesburgos, L., Mansson, E., Simonin, F., Kieffer, B., and Yu, L. 1998. Dynorphin A as a potential endogenous ligand for four members of the opioid receptor gene family. *J Pharmacol Exp Ther* 286:136–41.

Zhu, Y., and Pfaff, D. 1998. Differential regulation of AP-1 DNA binding activity in rat hypothalamus and pituitary by estrogen. *Brain Res Mol Brain Res* 55: 115–25.

Zhu, Y., Dellovade, T., and Pfaff, D. 1997a. Gender-specific induction of RNA synthesis in rat pituitary by estrogen and its interaction with thyroid hormones. *J Neuroendocrinol* 9:395–403.

———. 1997b. Interactions between hormonal and environmental signals on hypothalamic neurons: molecular mechanisms signaling environmental events. *Trends Endocrinol Metab* 8:111–15.

Zhu, Y., Ling, Q., Cai, L., Imperato-McGinley, J., and Pfaff, D. 1997. Regulation of preproenkephalin (PPE) gene expression by estrogen and its interaction with thyroid hormone. *Soc Neurosci Abstr* 23:798.

Zhu, Y.-S., and Pfaff, D. 1994. Protein-DNA binding assay for analysis of steroid-sensitive neurons in the mammalian brain. In *Neurobiology of Steroids: Methods in Neurosciences*, vol. 22, ed. E. De Kloet and W. Sutanto, 245. San Diego: Academic.

———. 1995. DNA binding of hypothalamic nuclear proteins on estrogen response element and preproenkephalin promoter: modification by estrogen. *Neuroendocrinology* 62:454–66.

Zhu, Y.-S., Cai, L.-Q., You, X., Imperato-McGinley, G., Chin, W., and Pfaff, D. 2000a. Receptor-isoform specific modification of estrogen induction of preproenkephalin gene expression by thyroid hormone. *Mol Brain Res.* In press.

Zhu, Y.-S., Koibuchi, N., Chin, W., and Pfaff, D. 2000b. Thyroid hormone receptor isoforms contributing to transcriptional actions opposing estrogen receptors. *Mol Brain Res.* In press.

Zhu, Y., Yen, P., Chin, W., and Pfaff, D. 1996. Estrogen and thyroid hormone interaction on regulation of gene expression. *Proc Natl Acad Sci USA* 93: 12587–92.

Zhuo, M., and Gebhart, G. 1991. Spinal serotonin receptors mediate descending facilitation of a nociceptive reflex from the nucleus gigantocellularis and gigantocellularis pars α in the rat. *Brain Res* 550:35.

———. 1992. Characterization of descending facilitation and inhibition of spinal nociceptive transmission from the nucleus reticularis gigantocellularis and gigantocellularis pars α in the rat. *J Neurophysiol* 67:1599.

Zhuo, M., and Gebhart, G.F. 1990. Spinal cholinergic and monoaminergic receptors mediate descending inhibition from the nuclei gigantocellularis and gigantocellularis pars α in the rat. *Brain Res* 535:67–78.

Ziegler, D., Cass, W., and Herman, J. 1999. Excitatory influence of the locus coeruleus in hypothalamic-pituitary adrenocortical axis responses to stress. *J Neuroendocrinol* 11:361–69.

Zieglgansberger, W., French, E.D., Siggins, G.R., and Bloom, F.E. 1979. Opioid peptides may excite hippocampal pyramidal neurons by inhibiting adjacent inhibitory interneurons. *Science* 205(4404):415–17.

Zukin, R.S., Eghbalai, M., Olive, D., Unterwald, E.M., and Tempel, A. 1988. Characterization and visualization of rat and guinea pig brain K opioid receptors: evidence for K-1 and K-2 opioid receptors. *Proc Natl Acad Sci USA* 85:4061–65.

# Index

Page numbers in *italics* denote figures; those followed by "t" denote tables.

251

# About the Authors

Richard J. Bodnar is professor and chair of psychology at Queens College, City University of New York, Flushing, New York.

Kathryn Commons is a postdoctoral fellow at the Rockefeller University, New York.

Donald W. Pfaff is professor of neurobiology and behavior at the Rockefeller University, New York.